INTRODUCING ISSUES WITH OPPOSING VIEWPOINTS®

AI, Robots, and the Future of the Human Race

Lisa Idzikowski, Book Editor

GREENHAVEN PUBLISHING

Published in 2020 by Greenhaven Publishing, LLC
353 3rd Avenue, Suite 255, New York, NY 10010

Copyright © 2020 by Greenhaven Publishing, LLC

First Edition

All rights reserved. No part of this book may be reproduced in any form without permission in writing from the publisher, except by a reviewer.

Articles in Greenhaven Publishing anthologies are often edited for length to meet page requirements. In addition, original titles of these works are changed to clearly present the main thesis and to explicitly indicate the author's opinion. Every effort is made to ensure that Greenhaven Publishing accurately reflects the original intent of the authors. Every effort has been made to trace the owners of the copyrighted material.

Library of Congress Cataloging-in-Publication Data

Names: Idzikowski, Lisa, editor.
Title: AI, robots, and the future of the human race / Lisa Idzikowski, book editor.
Other titles: Artificial intelligence, robots, and the future of the human race
Description: First edition. | New York : Greenhaven Publishing, 2020. | Series: Introducing issues with opposing viewpoints | Includes bibliographical references and index. | Audience: Grades 7–12.
Identifiers: LCCN 2019022830 | ISBN 9781534506596 (library binding) | ISBN 9781534506589 (paperback)
Subjects: LCSH: Artificial intelligence—Social aspects. | Robotics—Social aspects. | Artificial intelligence—Philosophy. | Robotics—Philosophy.
Classification: LCC Q334.7 .A43 2020 | DDC 303.48/3—dc23
LC record available at https://lccn.loc.gov/2019022830

Manufactured in the United States of America

Website: http://greenhavenpublishing.com

Contents

Foreword   5
Introduction   7

Chapter 1: What Is the Difference Between AI and Robots?

1. The Basics of Artificial Intelligence   12
   Steffen Herget
2. Artificial Intelligence: What Is It Really?   18
   Narasimha Prasanna
3. Can AI Achieve Consciousness?   23
   Vanessa Bates Ramirez
4. Industrial Robots Have Gone Through an Evolution in Their Development   30
   Alessandro Gasparetto and Lorenzo Scalera
5. Alan Turing's Test of Machine Intelligence Is Outdated   37
   Blaise Zerega

Chapter 2: Are AI and Robots Beneficial to Humans?

1. Can We Maintain Control Over AI?   43
   Vivian Michaels
2. Artificial Intelligence Could Threaten Humanity   48
   Rob Smith
3. Artificial Intelligence Will Benefit Society   54
   Kashyap Vyas
4. People Fear Artificial Intelligence   61
   Dave Dickinson
5. Artificial Intelligence Can Save Our Planet   67
   Celine Herweijer

Chapter 3: How Should AI and Robots Be Regulated to Benefit the Future of the Human Race?

1. Should Artificial Intelligence Be Regulated?   74
   Greg Benson
2. Will AI Make a World Without Work Possible?   81
   Ryan Avent
3. Will Robots Replace Human Employees?   87
   Owen Bowcott

4. Take Control of Autonomous Weapons Before It's Too Late   92
   Mattha Busby
5. Privacy Laws Need to Change in the Era of Big Data   98
   Danica Sergison
6. Should the Rights of Smart Robots Be Protected?   102
   E. Van Trigt

Facts About AI, Robots, and the Future of the Human Race   108
Organizations to Contact   110
For Further Reading   113
Index   117
Picture Credits   120

Foreword

Indulging in a wide spectrum of ideas, beliefs, and perspectives is a critical cornerstone of democracy. After all, it is often debates over differences of opinion, such as whether to legalize abortion, how to treat prisoners, or when to enact the death penalty, that shape our society and drive it forward. Such diversity of thought is frequently regarded as the hallmark of a healthy and civilized culture. As the Reverend Clifford Schutjer of the First Congregational Church in Mansfield, Ohio, declared in a 2001 sermon, "Surrounding oneself with only like-minded people, restricting what we listen to or read only to what we find agreeable is irresponsible. Refusing to entertain doubts once we make up our minds is a subtle but deadly form of arrogance." With this advice in mind, Introducing Issues with Opposing Viewpoints books aim to open readers' minds to the critically divergent views that comprise our world's most important debates.

Introducing Issues with Opposing Viewpoints simplifies for students the enormous and often overwhelming mass of material now available via print and electronic media. Collected in every volume is an array of opinions that captures the essence of a particular controversy or topic. Introducing Issues with Opposing Viewpoints books embody the spirit of nineteenth-century journalist Charles A. Dana's axiom: "Fight for your opinions, but do not believe that they contain the whole truth, or the only truth." Absorbing such contrasting opinions teaches students to analyze the strength of an argument and compare it to its opposition. From this process readers can inform and strengthen their own opinions, or be exposed to new information that will change their minds. Introducing Issues with Opposing Viewpoints is a mosaic of different voices. The authors are statesmen, pundits, academics, journalists, corporations, and ordinary people who have felt compelled to share their experiences and ideas in a public forum. Their words have been collected from newspapers, journals, books, speeches, interviews, and the Internet, the fastest growing body of opinionated material in the world.

Introducing Issues with Opposing Viewpoints shares many of the well-known features of its critically acclaimed parent series, Opposing

Viewpoints. The articles allow readers to absorb and compare divergent perspectives. Active reading questions preface each viewpoint, requiring the student to approach the material thoughtfully and carefully. Photographs, charts, and graphs supplement each article. A thorough introduction provides readers with crucial background on an issue. An annotated bibliography points the reader toward articles, books, and websites that contain additional information on the topic. An appendix of organizations to contact contains a wide variety of charities, nonprofit organizations, political groups, and private enterprises that each hold a position on the issue at hand. Finally, a comprehensive index allows readers to locate content quickly and efficiently.

Introducing Issues with Opposing Viewpoints is also significantly different from Opposing Viewpoints. As the series title implies, its presentation will help introduce students to the concept of opposing viewpoints and learn to use this material to aid in critical writing and debate. The series' four-color, accessible format makes the books attractive and inviting to readers of all levels. In addition, each viewpoint has been carefully edited to maximize a reader's understanding of the content. Short but thorough viewpoints capture the essence of an argument. A substantial, thought-provoking essay question placed at the end of each viewpoint asks the student to further investigate the issues raised in the viewpoint, compare and contrast two authors' arguments, or consider how one might go about forming an opinion on the topic at hand. Each viewpoint contains sidebars that include at-a-glance information and handy statistics. A Facts About section located in the back of the book further supplies students with relevant facts and figures.

Following in the tradition of the Opposing Viewpoints series, Greenhaven Publishing continues to provide readers with invaluable exposure to the controversial issues that shape our world. As John Stuart Mill once wrote: "The only way in which a human being can make some approach to knowing the whole of a subject is by hearing what can be said about it by persons of every variety of opinion and studying all modes in which it can be looked at by every character of mind. No wise man ever acquired his wisdom in any mode but this." It is to this principle that Introducing Issues with Opposing Viewpoints books are dedicated.

Introduction

"Success in creating effective AI could be the biggest event in the history of our civilization. Or the worst. We just don't know. So we cannot know if we will be infinitely helped by AI, or ignored by it and side-lined, or conceivably destroyed by it ... Unless we learn how to prepare for, and avoid, the potential risks, AI could be the worst event in the history of our civilization. It brings dangers, like powerful autonomous weapons, or new ways for the few to oppress the many. It could bring great disruption to our economy."

—Stephen Hawking, theoretical physicist

What comes to mind when people think about artificial intelligence (AI) and robots? Do they recall how R2-D2 and C-3PO helped Luke Skywalker, Princess Leia, and Han Solo fight against the Empire in *Star Wars*, or picture the Iron Giant befriending a lonely boy and saving his town in *The Iron Giant*? What about big, soft Baymax in *Big Hero 6*, who exists to help care for people? These images put a positive spin on AI and robots and perhaps ease the worries of some individuals.

But not all cultural renderings of AI and robots portray them as friendly to humans. As the name implies, in the movie *Terminator* the T-800 Terminator destroyed nearly any human in sight because it was given orders to kill one particular woman and her unborn son. Then there's Megatron, the brutal robotic warlord from *Transformers*. And what about the very famous HAL 9000, the mutinous machine from *2001: A Space Odyssey*? If these depictions represent the potential future of AI and robots, it's no wonder that some people are ill at ease about the topic.

Interestingly, the term "artificial intelligence" (AI) was first used in 1956 by a group of researchers coming together to talk about the topic of thinking machines. John McCarthy wanted a name that made the idea of thinking machines seem neutral instead of negative or positive.[1] Today, there are different ways of defining AI. *Merriam-Webster* defines artificial intelligence as "a branch of computer science dealing with the simulation of intelligent behavior in computers" and "the capability of a machine to imitate intelligent human behavior."[2] The *Encyclopedia Britannica* defines AI as "the ability of a digital computer or computer-controlled robot to perform tasks commonly associated with intelligent beings."[3] Intelligent beings are assumed to be those that are able to adapt to changing circumstances.

As the introductory quote from Stephen Hawking suggests, AI and robots could be one of the best or worst inventions that humankind has ever developed. A well-regarded and world-renowned physicist, Hawking was also an outspoken critic of AI. He periodically shared his thoughts and worries about the subject. The entrepreneur and Space-X engineer and CEO Elon Musk is another person who is sounding the alarm about the possible negative outcomes he foresees when sentient AI takes over and turns on humans.

However, on the other side of the debate are individuals who believe that researchers and scientists will be able to stay ahead of these creations and remain in control. Steve Wozniak, co-founder of Apple Computers, used to be in the same camp as Hawking and Musk. Now he says "artificial intelligence doesn't scare me at all."[4] Wozniak believes that the thinking ability of machines is eons behind the capability of a young child, and that it would take hundreds of years for it to catch up.

Even so, there are other issues that are important in the debate surrounding AI and robots. A big and commonly referenced concern is the notion that robots will take away many of the jobs currently performed by people. If this happens, what will individuals do to obtain the money they currently earn from working? How will they pay for things? What will they do with all the extra time if they're not working? Many people in the US are worried about this. In a 2017 Pew Research survey, 72 percent of those surveyed said they are worried about this happening, and 25 percent say they are very worried. On the flip side, only 33 percent are enthusiastic about robots taking over human tasks. Even if the government would give people a living income if their jobs were taken over by robots, only

30 percent favored this option.[5] The message is clear: people want to keep their jobs!

Ethical and legal questions abound when considering the future of AI. What about the issue of privacy in this time of big data? Most people have a presence on social media, which gives companies and other entities access to large amounts of personal information. How should this be controlled? Algorithms are being used on this personal data to determine what information is directed at each user, raising concerns about whether this use of AI is ethical and good for the public. Another issue that worries some people is the future capabilities of autonomous weapons, which are those that operate without human intervention. Proponents generally point to the fact that this type of weaponry will save human lives through taking them out of risky situations. Opponents insist that smart weapons could essentially turn on humans and there would be no way to stop them and the extensive killing that would ensue. The thought that smart robots will someday have a physical body, developed senses, and the ability to reason and manipulate their environment underlies another interesting controversy. Some experts believe that intelligent robots will someday need to be provided "rights." If people and animals deserve to have rights because they possess these qualities, their reasoning goes, then so do intelligent robots.

Clearly, the advancement of AI and robots and how these forms of technology affect the human race are controversial concerns. Though they are machines that people created in the hope of benefitting the human race, they can also be viewed as a possible mistake or serious threat to humanity if their development gets out of hand. The current debate that surrounds this interesting topic is explored in *Introducing Issues with Opposing Viewpoints: AI, Robots, and the Future of the Human Race*, and sheds light on this divisive and ongoing contemporary issue.

Notes

1. Martin Childs, "John McCarthy: Computer Scientist Known as the Father of AI," *Independent*, November 1, 2011. https://www.independent.co.uk/news/obituaries/john-mccarthy-computer-scientist-known-as-the-father-of-ai-6255307.html.
2. "Artificial Intelligence," *Merriam-Webster*. https://www.merriam-webster.com/dictionary/artificial%20intelligence.

3. "Artificial Intelligence," *Encyclopedia Britannica*, May 9, 2019. https://www.britannica.com/technology/artificial-intelligence.
4. Catherine Clifford, "Steve Wozniak Explains Why He Used to Agree with Elon Musk, Stephen Hawking on A.I.—But Now He Doesn't," *CNBC Make It*, February 23, 2018. https://www.cnbc.com/2018/02/23/steve-wozniak-doesnt-agree-with-elon-musk-stephen-hawking-on-a-i.html.
5. Aaron Smith and Monica Anderson, "Americans' Attitudes Toward a Future in Which Robots and Computers Can Do Many Human Jobs," Pew Research Center, October 4, 2017. https://www.pewinternet.org/2017/10/04/americans-attitudes-toward-a-future-in-which-robots-and-computers-can-do-many-human-jobs/.

Chapter 1

What Is the Difference Between AI and Robots?

Sophia (pictured) is a humanoid robot that was created in 2016.

Viewpoint 1

The Basics of Artificial Intelligence

Steffen Herget

"AI isn't something that just came out of nowhere recently."

In the following viewpoint, Steffen Herget explains AI, its history, and its basic definitions and applications. According to Herget, AI as we know it has been in development since the 1950s, and today many people already interact with these systems when they talk with a customer service representative or use a home assistant. He describes the various types of AI and the tools and methods used for applying AI to daily life, explaining the distinctions between symbolic and neural AI, as well as machine learning and deep learning. Steffen Herget is a senior editor for AndroidPIT International. He is based in Berlin.

AS YOU READ, CONSIDER THE FOLLOWING QUESTIONS:

1. How does Herget define AI?
2. What systems have people probably come into contact with, as suggested by the author?
3. As indicated in this viewpoint, which AI project appeared on TV?

"What Is AI? History, Definitions and Applications," by Steffen Herget, Androidpit, October 16, 2017. Reprinted with permission.

Everyone is talking about artificial intelligence, also known in its abbreviated form, AI. But what is it all about? That's precisely what we'll be explaining today.

History

Artificial intelligence is increasingly playing a greater role in our lives, and the latest trend are AI chips and the accompanying smartphone applications. But this technology began to be developed as early as in the 50s with the Dartmouth Summer Research Project on Artificial Intelligence at Dartmouth College in the US Its origins date back even further to the work of Alan Turing—to whom we can attribute the famous Turing test—Allen Newell and Herbert A. Simon, but AI did not make it into the spotlight on the world stage until the arrival of chess supercomputer Deep Blue by IBM, which was the first machine to defeat the then-defending world chess champion Garry Kasparov in a match in 1996. AI algorithms have been used in data centers and on large computers for many years, but is only more recently present in the realm of consumer electronics.

Definition of Artificial Intelligence

The definition of artificial intelligence characterizes it as a branch of computer science that deals with automating intelligent behavior. Here's the hard part: Since you cannot precisely define intelligence *per se*, artificial intelligence cannot be exactly defined either. Generally speaking, the term is used to describe systems whose objective is to use machines to emulate and simulate human intelligence and the corresponding behavior. This can be accomplished with simple algorithms and pre-defined patterns, but can become far more complex as well.

Various Kinds of AI

Symbolic or symbol-manipulating AI works with abstract symbols that are used to represent knowledge. It is the classic AI that pursues the idea that human thinking can be reconstructed on a hierarchical, logical level. Information is processed from above, working with human-readable symbols, abstract connections and logical conclusions.

Watson (pictured above at center) is a question-answering computer system that competed on the game show Jeopardy! *in 2011, defeating champions Ken Jennings (left) and Brad Rutter (right).*

Neural AI became popular in computer science in the late 80s. Here, knowledge is not represented through symbols, but rather artificial neurons and their connections—sort of like a reconstructed brain. The gathered knowledge is broken down into small pieces—the neurons—and then connected and built into groups. This approach is known as the bottom-up method that works its way from below. Unlike symbolic AI, a neural system must be trained and stimulated so that the neural networks can gather experience and grow, therefore accumulating greater knowledge.

Neural networks are organized into layers that are connected to each other via simulated lines. The uppermost layer is the input layer, which works like a sensor that accepts the information to be processed and passes it on below. This is now followed by at least two—or more than twenty in large systems—layers that are hierarchically above each other and that send and classify information via the connections. At the very bottom is the output layer, which generally has the least number of artificial neurons. It provides the calculated data in a machine-readable form, i.e. picture of a dog during the day with a red car.

Methods and Tools

There are various tools and methods for applying artificial intelligence to real-world scenarios, some of which can be used in parallel.

The foundation of all this is machine learning, which is defined as a system that builds up knowledge from experience. This process gives the system the ability to detect patterns and laws—and with ever-increasing speed and accuracy. In machine learning, both symbolic and neural AI is used.

Deep learning is a subtype of machine learning that is becoming ever more important. Only neural AI, i.e. neural networks, are used in this case. Deep learning is the foundation for most current AI applications. Thanks to the possibility of increasingly expanding the design of the neural networks and making them more complex and powerful with new layers, deep learning is easily scalable and adaptable to many applications.

There are three learning processes for training neural networks: supervised, non-supervised and reinforcement learning, providing many different ways to regulate how an input becomes the desired output. While target values and parameters are specified from the outside in supervised learning, in unsupervised learning, the system attempts to identify patterns in the input that have an identifiable structure and can be reproduced. In reinforcement learning, the machine also works independently, but is rewarded or punished depending on the success or failure.

Applications

Artificial intelligence is already being used in many areas, but by no means are all of them visible at first glance. Therefore, selecting scenarios that take advantage of the possibilities of this technology is by no means a completed list.

Artificial intelligence's mechanisms are excellent for detecting, identifying, and classifying objects and persons on pictures and videos. To that end, simple but CPU-intensive pattern detection is used. If the image information is decrypted and machine-readable in the first place, photos and videos can be easily divided into categories, searched and found. Such recognition is also possible for audio data.

> **FAST FACT**
> The first commercial application for IBM's Watson was in 2013, when it was put to use diagnosing lung cancer.

Customer service is increasingly using chatbots. These text-based assistants perform recognition using key words that the customer may tell it and they respond accordingly. Depending on the use, this assistant can be more or less complex.

Opinion analysis is not only used for forecasting elections in politics, but also in marketing and many other areas. Opinion mining, also known as sentiment analysis, is used to scour the internet for opinion and emotional expressions, allowing for the creation of a largely anonymized opinion survey.

Search algorithms like Google's are naturally top secret. The way in which search results are calculated, measured and outputted are largely determined by mechanisms that work with machine learning.

Word processing, or checking the grammar and spelling of a text, is a classic application of symbolic AI that has been used for a long time. Language is defined as a complex network of rules and instructions that analyzes blocks of text in a sentence and, under some circumstances, can identify and correct errors.

These abilities are also used in synthesizing speech, which is currently the talk of the town with assistant systems like Siri, Cortana, Alexa or Google Assistant.

On new smartphone chips like the Kirin 970, artificial intelligence is integrated into its own component, the NPU or neural processing unit. The processor is making its debut in the Huawei Mate 10. You will learn more about it and the roles that the technology will play on the Huawei smartphone once we have a chance to experiment with it in the near future. Qualcomm has already been working on an NPU, the Zeroth processor, for two years, and the new Apple A11 chip contains a similar component.

Furthermore, there are numerous research projects on artificial intelligence and the most prominent of all may be IBM's Watson. The computer program had already made its first public debut in 2011 on the quiz show *Jeopardy*, where it faced off against two human candidates. Watson won, of course, and additional publicity appearances took place afterwards. A Japanese insurance company has been using Watson since January to check insured customers, their history and medical data and to evaluate injuries and illnesses. According to the company's information, Watson has replaced roughly 30 employees. Loss of jobs through automation is just one of the ethical and social issues surrounding AI that is the subject of corporate and academic research.

Projection

AI isn't something that just came out of nowhere recently, but it is coming close to a breakthrough in the world of consumer electronics, which is more than enough reason for everyone to keep up to date with this topic in the future.

> **EVALUATING THE AUTHOR'S ARGUMENTS:**
>
> At the end of this viewpoint, Steffen Herget makes the argument that it is important for any consumer of electronics to have a background understanding of AI and its history. Do you think it is necessary to have this knowledge in order to use and appreciate technology?

Viewpoint 2

Artificial Intelligence: What Is It Really?

Narasimha Prasanna

> "AGI (Artificial General Intelligence) is a term used to describe real intelligent systems."

In the following excerpted viewpoint, Narasimha Prasanna contends that AI has been around a long time, but each generation of AI has had its limits. Prasanna argues that AI systems are not actually intelligent and examines the example of machine learning called "supervised learning." Through the use of supervised learning and similar systems, it is possible to create the illusion of real intelligence by imitating human thought processes, but AI "learning" has limitations that stand in the way of true intelligence. Prasanna also delves into reinforcement learning, which many scientists believe will be the way to approach artificial general intelligence (AGI). Narasimha Prasanna is a programmer who codes in Python and JavaScript.

"What is Artificial General Intelligence?" by Narasimha Prasanna HN, February 12, 2018. Reprinted with permission.

AS YOU READ, CONSIDER THE FOLLOWING QUESTIONS:
1. What are the limitations of supervised learning listed in this viewpoint?
2. What allows machines to learn from examples much like the human brain, according to Prasanna?
3. According to the author, what is AGI?

Artificial Intelligence is a branch of Computer Science (or Science) which deals with the creation of intelligent systems. Intelligent systems are those systems which possess intelligence just like humans.

The science of artificial intelligence is not new, the term artificial intelligence has been mentioned in manuscripts of Ancient Greece and Egypt. Greeks believed in [the] god Hephaestus, also known as [the] God of Blacksmiths, according to a Greek mythology Hephaestus made intelligent weapons for all Gods, in their view, the goal of artificial intelligence is to: be helpful for people to achieve a certain goal, be able to operate automatically and be programmed in advance to react in different ways depending on the situation.

Well, the term Artificial Intelligence has become popular in the field of Entertainment, we can see lots of movies based on the concept of super intelligence. (*Ex-Machina, Her, AI, Avengers—Age of Ultron* etc.). But artificially intelligent systems what we see today are in no match with so called "super intelligent" systems.

Difference Between Real AI and AI Systems of Today

Artificial intelligence, as stated earlier is not a new field, many philosophers and scientists had some imagination about AI since humanity, but they were all limited by the technologies of their time. Today with the presence of powerful super computers, we're able to build AI systems which serves the required purpose. But, are they really intelligent? The answer is: No, they are not. Let's see how.

With the help of computers and availability of large enough datasets on the internet, the so called Machine Learning came into [the]

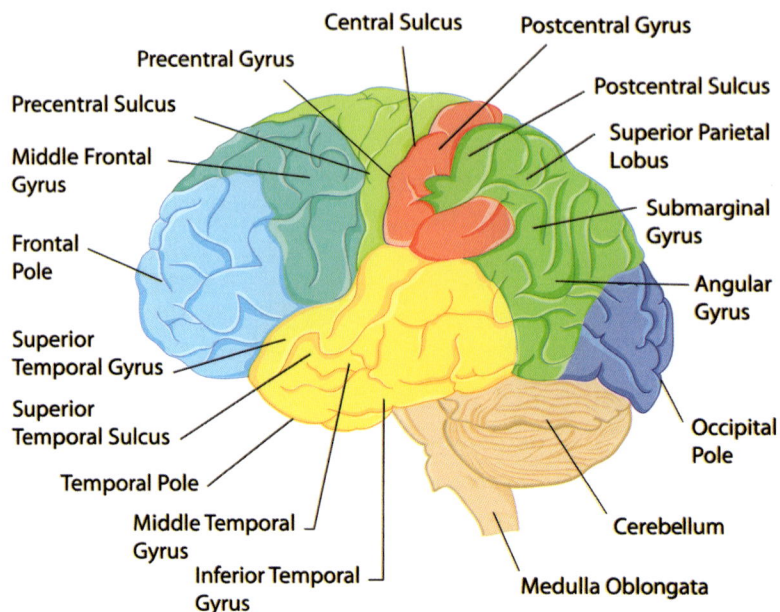

In order for AI to be able to truly "learn," some argue that it must function the same way as the human brain.

picture. Machine Learning provides a set of mathematical concepts using AI that can be implemented in the real world.

Neural networks roughly simulate the working of the human brain to make machines learn from examples. Deep learning has helped many tech giants like Google and Apple to improve their products economically by implementing so many new buzzy tech like face recognition, language understanding, image understanding, etc. But what you think of as so-called deep learning isn't real intelligence. The field of Machine Learning which requires enormous data sets to learn either to classify objects or to make predictions is called Supervised Learning.

The so-called Supervised Learning creates an illusion of Intelligence, but at its core it is just a mathematical optimization. Even though it possesses an ability to make decisions and classify datasets, it is very narrow in the way it works.

We are very familiar with techniques to create supervised learning systems. Given a large dataset, Supervised Learning systems learn mapping between inputs and outputs, and thus it can predict outputs for unseen inputs. But this is not what our brain actually does, our brain doesn't require 10,000 images of cats to recognize a cat, and even our brain can do so many things which Supervised Learning systems cannot.

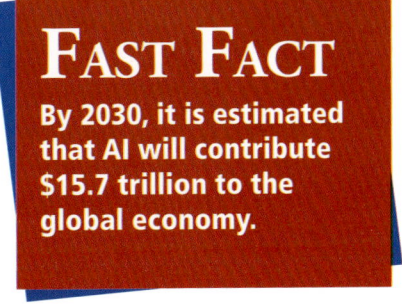

FAST FACT
By 2030, it is estimated that AI will contribute $15.7 trillion to the global economy.

Limitations of Supervised Learning

Even though Supervised Learning can be used to create so many amazing things, it has so many limitations:

- Its thinking is always limited to a specific domain.
- Its intelligence depends on the training dataset you have used. Again, you are in control.
- It cannot be used in environments that dynamically change.
- Can be used only for either classification or regression, but not for control problems.
- It requires huge datasets, if not it lacks accuracy. Obtaining datasets can be a problem.

What is AGI?

AGI (Artificial General Intelligence) is a term used to describe real intelligent systems. Real intelligent systems possess ability to think generally, to make decisions irrespective of any previous training. Here decisions are made based on what they've learnt on their own. It can be really difficult to design such systems as technology of today is somewhat limited, but we can create so called "Partial AGI."

Reinforcement Learning

Even today, many scientists believe in Reinforcement Learning as a way to achieve so called AGI. Reinforcement Learning can be applied as a solution for problems which Supervised Learning cannot solve. Let us take a simple example to understand one of the basic tasks of walking. Walking is a human task which we are very natural and good at. A baby learns to walk by itself, without having to search for a dataset—a human brain can do so by learning from mistakes. But once it becomes perfect, it can walk thousands of steps, even a single mismatch in step size can be identified by it. Steps taken by brain are always optimal, the speed (or simply step-size) taken is always in such a way that the destination should be reached in minimal time in the same way, the energy taken for each step should be minimal. So the rate of walking depends on energy as well as how fast you have to reach the destination. The same problem of walking can be applied to so many areas like robotics where locomotion is necessary. Of course, as stated earlier, Supervised Learning cannot do that.

[…]

EVALUATING THE AUTHOR'S ARGUMENTS:

In this viewpoint Narasimha Prasanna compares supervised learning to artificial general intelligence. Do the arguments presented in the viewpoint suggest that AGI is possible? Why or why not? How does supervised learning differ from the way humans learn?

Viewpoint 3

Can AI Achieve Consciousness?

Vanessa Bates Ramirez

"The biggest question that arises here, and one that's become a popular theme across stories and films, is if machines achieve human-level general intelligence, does that also mean they'd be conscious?"

In this viewpoint, Vanessa Bates Ramirez responds to Ray Kurzweil's often-cited prediction that artificial intelligence will soon be able to pass for humans and will possess the same amount of emotional intelligence. She has her doubts about whether this will be achievable in the next ten years (as Kurzweil predicted) or even in the foreseeable future. She argues that because we don't fully understand how the brain works and what consciousness is, and because AI is modeled on the human brain and aims to achieve consciousness, it will be impossible to make these significant advances in AI without first answering these other questions. However, she asserts that advancements in AI are still occurring despite this. Vanessa Bates Ramirez is senior editor at *Singularity Hub*, which is part of Singularity University and aims at promoting understanding of the breakthroughs and future trends in exponential technologies.

"What Would It Mean for AI to Become Conscious?" by Vanessa Bates Ramirez, Singularity Hub, March 26, 2019, https://singularityhub.com/2019/03/26/what-would-it-mean-for-ai-to-become-conscious/. Licensed under CC BY-ND 4.0.

AS YOU READ, CONSIDER THE FOLLOWING QUESTIONS:
1. Why does the author think it's strange that we model AI on the human brain?
2. What are some of the examples of advances in brain-machine interfaces in the medical field provided in the viewpoint?
3. What definition of AI does the author offer in this viewpoint?

As artificial intelligence systems take on more tasks and solve more problems, it's hard to say which is rising faster: our interest in them or our fear of them. Futurist Ray Kurzweil famously predicted that "By 2029, computers will have emotional intelligence and be convincing as people."

We don't know how accurate this prediction will turn out to be. Even if it takes more than 10 years, though, is it really possible for machines to become conscious? If the machines Kurzweil describes say they're conscious, does that mean they actually are?

Perhaps a more relevant question at this juncture is: what *is* consciousness, and how do we replicate it if we don't understand it?

In a panel discussion at South By Southwest titled "How AI Will Design the Human Future," experts from academia and industry discussed these questions and more.

Wait, What Is AI?

Most of AI's recent feats—diagnosing illnesses, participating in debate, writing realistic text—involve machine learning, which uses statistics to find patterns in large datasets then uses those patterns to make predictions. However, "AI" has been used to refer to everything from basic software automation and algorithms to advanced machine learning and deep learning.

"The term 'artificial intelligence' is thrown around constantly and often incorrectly," said Jennifer Strong, a reporter at the *Wall Street Journal* and host of the podcast "The Future of Everything." Indeed, one study found that 40 percent of European companies that claim to be working on or using AI don't actually use it at all.

Ray Kurzweil (pictured) is an American inventor and futurist who has written extensively on the topic of the singularity.

Dr. Peter Stone, associate chair of computer science at UT Austin, was the study panel chair on the 2016 One Hundred Year Study on Artificial Intelligence(or AI100) report. Based out of Stanford University, AI100 is studying and anticipating how AI will impact our work, our cities, and our lives.

"One of the first things we had to do was define AI," Stone said. They defined it as a collection of different technologies inspired by the human brain to be able to perceive their surrounding environment and figure out what actions to take given these inputs.

Modeling on the Unknown

Here's the crazy thing about that definition (and about AI itself): we're essentially trying to re-create the abilities of the human brain without having anything close to a thorough understanding of how the human brain works.

"We're starting to pair our brains with computers, but brains don't understand computers and computers don't understand brains," Stone said. Dr. Heather Berlin, cognitive neuroscientist and professor of psychiatry at the Icahn School of Medicine at Mount Sinai, agreed. "It's still one of the greatest mysteries how this three-pound piece of matter can give us all our subjective experiences, thoughts, and emotions," she said.

This isn't to say we're not making progress; there have been significant neuroscience breakthroughs in recent years. "This has been the stuff of science fiction for a long time, but now there's active work being done in this area," said Amir Husain, CEO and founder of Austin-based AI company Spark Cognition.

Advances in brain-machine interfaces show just how much more we understand the brain now than we did even a few years ago. Neural implants are being used to restore communication or movement capabilities in people who've been impaired by injury or illness. Scientists have been able to transfer signals from the brain to prosthetic limbs and stimulate specific circuits in the brain to treat conditions like Parkinson's, PTSD, and depression.

But much of the brain's inner workings remain a deep, dark mystery—one that will have to be further solved if we're ever to get from narrow AI, which refers to systems that can perform specific tasks and is where the technology stands today, to artificial general intelligence, or systems that possess the same intelligence level and learning capabilities as humans.

The biggest question that arises here, and one that's become a popular theme across stories and films, is if machines achieve human-level general intelligence, does that also mean they'd be conscious?

Wait, What Is Consciousness?

As valuable as the knowledge we've accumulated about the brain is, it seems like nothing more than a collection of disparate facts when we try to put it all together to understand consciousness.

> **FAST FACT**
> Moore's Law claims that every two years the speed and capability of computers will double while the price of them will be cut in half.

"If you can replace one neuron with a silicon chip that can do the same function, then replace another neuron, and another—at what point are you still you?" Berlin asked. "These systems will be able to pass the Turing test, so we're going to need another concept of how to measure consciousness."

Is consciousness a measurable phenomenon, though? Rather than progressing by degrees or moving through some gray area, isn't it pretty black and white—a being is either conscious or it isn't?

This may be an outmoded way of thinking, according to Berlin. "It used to be that only philosophers could study consciousness, but now we can study it from a scientific perspective," she said. "We can measure changes in neural pathways. It's subjective, but depends on reportability."

She described three levels of consciousness: pure subjective experience ("Look, the sky is blue"), awareness of one's own subjective experience ("Oh, it's me that's seeing the blue sky"), and relating one subjective experience to another ("The blue sky reminds me of a blue ocean").

"These subjective states exist all the way down the animal kingdom. As humans we have a sense of self that gives us another depth to that experience, but it's not necessary for pure sensation," Berlin said.

Husain took this definition a few steps farther. "It's this self-awareness, this idea that I exist separate from everything else and that I can model myself," he said. "Human brains have a wonderful simulator. They can propose a course of action virtually, in their minds, and see how things play out. The ability to include yourself as an actor means you're running a computation on the idea of yourself."

Most of the decisions we make involve envisioning different outcomes, thinking about how each outcome would affect us, and choosing which outcome we'd most prefer.

"Complex tasks you want to achieve in the world are tied to your ability to foresee the future, at least based on some mental model," Husain said. "With that view, I as an AI practitioner don't see a problem implementing that type of consciousness."

Moving Forward Cautiously (But Not Too Cautiously)

To be clear, we're nowhere near machines achieving artificial general intelligence or consciousness, and whether a "conscious machine" is possible—not to mention necessary or desirable—is still very much up for debate.

As machine intelligence continues to advance, though, we'll need to walk the line between progress and risk management carefully.

Improving the transparency and explainability of AI systems is one crucial goal AI developers and researchers are zeroing in on. Especially in applications that could mean the difference between life and death, AI shouldn't advance without people being able to trace how it's making decisions and reaching conclusions.

Medicine is a prime example. "There are already advances that could save lives, but they're not being used because they're not trusted by doctors and nurses," said Stone. "We need to make sure there's transparency." Demanding too much transparency would also be a mistake, though, because it will hinder the development of systems that could at best save lives and at worst improve efficiency and free up doctors to have more face time with patients.

Similarly, self-driving cars have great potential to reduce deaths from traffic fatalities. But even though humans cause thousands of deadly crashes every day, we're terrified by the idea of self-driving cars that are anything less than perfect. "If we only accept autonomous cars when there's zero probability of an accident, then we will never accept them," Stone said. "Yet we give 16-year-olds the chance to take a road test with no idea what's going on in their brains."

This brings us back to the fact that, in building tech modeled after the human brain—which has evolved over millions of years—we're working towards an end whose means we don't fully comprehend, be it something as basic as choosing when to brake or accelerate or something as complex as measuring consciousness.

"We shouldn't charge ahead and do things just because we can," Stone said. "The technology can be very powerful, which is exciting, but we have to consider its implications."

> **EVALUATING THE AUTHOR'S ARGUMENTS:**
>
> Does the author believe that we should wait until all risks related to AI are eliminated before allowing it to develop? Why or why not? According to the evidence and arguments presented in the viewpoint, do you think the potential benefits of AI outweigh the risks?

Viewpoint 4

Industrial Robots Have Gone Through an Evolution in Their Development

"The evolution of industrial robotics is not over…"

Alessandro Gasparetto and Lorenzo Scalera

In the following excerpted viewpoint, Gasparetto and Scalera outline the development of industrial robots. The authors analyze each of the four generations of robots and the improvements made by each generation. In addition to the technological factors at play, the authors outline the social, economic, and political factors that affected innovations in robotics, suggesting that the relationships between these spheres has long impacted the advancement of robotic technology. The authors conclude that industrial robots will continue becoming more robust and valuable to factory automation. Alessandro Gasparetto and Lorenzo Scalera are affiliated with the Polytechnic Department of Engineering and Architecture at the University of Udine in Italy.

"A Brief History of Industrial Robotics in the 20th Century," by Alessandro Gasparetto and Lorenzo Scalera, Advances in Historical Studies, February 15, 2019, https://file.scirp.org/Html/2-2810274_90517.htm. Licensed under CC BY 4.0 International

AS YOU READ, CONSIDER THE FOLLOWING QUESTIONS:
1. According to the authors, how many generations of industrial robots have there been?
2. What drawback did the second generation of robots have, as reported in the viewpoint?
3. What higher-level intelligence do fourth generation robots possess, according to Gasparetto and Scalera?

The idea to design and build some sort of beings, or devices that could carry out repetitive or heavy tasks, thus relieving men from this burden, dates back to ancient times. Since the Greek-Hellenistic age some of these devices, which were named automata, have been designed and created by ingenious inventors, belonging to several different civilizations through the centuries. The term "automata" mainly refers to human-like devices, while the term "robot" has a more general meaning.

The origin of the term "robot" is placed in more recent times: namely, it comes from the Czech word "robota," meaning "heavy work" or "forced labour." The introduction of this term is due to the Czech writer Karel Čapek (1890–1938), who used it for the first time in 1920 in his novel "R.U.R.: Rossum's Universal Robots."

On the other hand, the word "Robotics" was employed for the first time by Isaac Asimov (1920–1992) in his novel "Runaround" (1942), contained in the famous series "I, Robot." In that novel he defined three rules concerning the behavior of robots and the interaction with humans: these rules would later be named the three Laws of Robotics.

In the literature, there are not so many works dealing with Robotics from a historical perspective. For instance, Ceccarelli dealt with this topic, while Gasparetto (2016) presents a historical outline of robotics from ancient times until the Industrial Revolution.

Industrial applications of Robotics gained a paramount importance in the last century. The beginning of "Industrial Robotics," as we currently define it, can be dated back to the 1950's, although some kinds of automatization in the industrial environment started to appear since the times of the Industrial Revolution.

Robots are used for various industrial purposes, including the manufacturing of cars. Their industrial abilities and applications have developed significantly over the past century.

In this paper, the main milestones of the history of industrial robotics, from its beginning (in the 1950's and even earlier) to the end of the 20th century, will be mentioned and described.

The evolution of industrial robots can be subdivided in four categories, the first three covering the timespan from the 1950's to the end of the 1990's. The robots of the fourth generation (which ranges from 2000 to nowadays), that are characterized by high-level "intelligent" features (such as the capability of performing advanced computations, logical reasoning, deep learning, complex strategies, collaborative behavior), are not analyzed in this paper.

In the scientific literature, not so many works on the history of industrial robots are present. In some books there are some sketches on this topic, which could also be found in some internal reports.

The First Generation of Industrial Robots (1950–1967)

Some authors proposed a chronological categorization of industrial robots, by defining four "generations."

The first generation of industrial robot spans from 1950 to 1967. The robots of this generation were basically programmable machines that did not have the ability to really control the modality of task execution; moreover, they had no communication with the external environment. With respect to the hardware, the first generation robots were provided with low-tech equipment, and servo-controllers were not present. A peculiar feature of these robots is the strong noise they produced, when their arms collided with the mechanical stops built to limit the movement of the axes. With respect to the actuators, almost all the first generation robots employed pneumatic actuators and were controlled by a sort of "logic gates" acting as automatic regulators. Such "logic gates" were basically cams which activated pneumatic valves, or relays which controlled solenoid valves. Finally, the tasks that the first generation robots were capable to perform were necessarily quite trivial, such as loading-unloading or simple material handling operations.

[…]

The Second Generation of Industrial Robots (1968–1977)

The industrial robots of the second generation (conventionally ranging from 1968 to 1977) were basic programmable machines with limited possibilities of self-adaptive behavior and elementary capabilities to recognize the external environment. These robots used servo-controllers, which enabled them to perform both point-to-point motion, and continuous paths as well. Their control system consisted of microprocessors or of Programmable Logic Controllers (PLC), and they could be also programmed by an operator by means of a teach box. With respect to those belonging to the first generation, these robots could carry out more complex tasks (e.g. control of work centers). However, their level of versatility was not very high, because each robot had its own software, which was dedicated to a specific task. Hence, these robots turned out to be application-specific devices, meaning that it was very difficult to employ the same robot for different tasks, since this would require a substantial modification of the controller, and a thorough reprogramming of the operating software. With respect to diagnostics, the robots of the second generation were not very performant, since the only diagnostic reports they could produce were those related to failures, which were reported by means of indicator lights, without any hint related to the cause of the failure that was left to the operator to trace.

[…]

The economic and geopolitical situation, at the international level, also pushed the trend towards electrically driven robots: for instance, the price of crude oil suddenly grew after the oil crisis following the Kippur war (October 1973). The companies were thus forced to find more efficient ways of production: robot (and in particular electrically driven robots) was consistent with the goals of reducing the costs and increasing the productivity. This gave a boost to the installations of industrial robots, which increased more than 30% per year in the second half of the 1970's.

[…]

The Third Generation of Industrial Robots (1978–1999)

The industrial robots of the third generation (conventionally ranging from 1978 to 1999) were characterized by a larger extent of interaction with both the operator and the environment, through some kind of complex interfaces (such as vision or voice). They also had some self-programming capabilities, and could reprogram themselves, although by a little amount, in order to execute different tasks. These robots were provided with servo controls, and could execute complex tasks, by moving either from point to point or along continuous paths. They could be programmed either on-line (the operator could use a teach box with a keyboard) or off-line, being connected to a PLC or a PC, which allowed to use a high-level language for motion programming and enabled the robots to be interfaced with a CAD or a database. The possibility of high-level, off-line programming enlarged the operational potential of the robots: for in-stance, they could elaborate data from sensor reading, in order to adjust the robot movements taking into account changes in the environment (e.g. changes in position and orientation of the workpieces). Moreover, the diagnostic capabilities could be greatly enhanced: these robots could produce not only an indication of failure detection, but also a report on the location and on the type of the failure. In addition, some sort of "intelligence" was present in the robots of the third generation, with some (although limited) adaptive capabilities. These capabilities could be employed in some more complex tasks (such as tactile inspection, assembly operations, arc welding), by using the data coming from vision or perception systems to locate the objects and the workpieces and guide the joint movements according to the task to be performed, taking into account the possibility of small changes in the position of the objects.

> **Fast Fact**
> A type of modern robot called a collaborative robot—or "cobot"—is meant to work alongside humans to help complete work.

[…]

The end of the third generation is conventionally set to the end of the century; beginning from the year 2000, the industrial robots are considered to belong to the fourth generation (which extends up to the current days). Such robots feature high-level "intelligent" capabilities (such as performing advanced computations, logical reasoning, deep learning, complex strategies, collaborative behavior).

Conclusion

In this paper, a brief history of industrial robotics in the 20th century was presented. The evolution of the industrial robots was conventionally categorized into four generations, of which the first three cover the time-span from the 1950's to the end of the century.

In this historical sketch, not only the scientific and technical evolution was taken into account, but some considerations about the economic and geopolitical issues that determined the diffusion of industrial robots, were also done.

The evolution of industrial robotics is not over, but is still developing in the current days: innovative ideas and novel hardware devices, together with some new programming techniques connected with Artificial Intelligence, are revolutionizing the concept of industrial automation and giving a new youth to the factory environment.

> **EVALUATING THE AUTHORS' ARGUMENTS:**
>
> In this viewpoint Alessandro Gasparetto and Lorenzo Scalera provide a look at how industrial robots developed from the 1950s to the present and examine how social, political, and economic factors impacted these advancements. Considering some of the major events and challenges of today's world, what do you predict could be some of the applications for robotics in the twenty-first century? Explain your reasoning.

Viewpoint 5

Alan Turing's Test of Machine Intelligence Is Outdated

"Advances like self-driving cars, speech processing, and image recognition have rendered the test less relevant."

Blaise Zerega

In the following viewpoint, Blaise Zerega provides an introduction to the computer scientist Alan Turing, known for developing the Turing test for machine intelligence. Zerega outlines the thoughts of several computer experts on the pros and cons of Turing's test and what it means for future attempts at determining the intelligence of AI systems. Some assert that although the test once served a useful purpose in encouraging the scientific community to create AI that matches human intelligence, AI now exceeds the capabilities of humanity, and so it should have a new test that matches its abilities. Blaise Zerega is the editor-in-chief for the tech company All Turtles.

"The Turing test is tired. It's time for AI to move on," by Blaise Zerega, All Turtles Corporation, September 20, 2017. Reprinted with permission.

AS YOU READ, CONSIDER THE FOLLOWING QUESTIONS:
1. As explained in the viewpoint, who was Alan Turing?
2. According to Zerega, is there a suitable alternative to the Turing test?
3. What is the consensus of the experts interviewed regarding their attitudes toward the Turing test?

This weekend at England's famed Bletchley Park, a bit of software fooled some humans into thinking it was, well, human. Mitsuku, an animated chatbot that calls itself "an artificial life form living on the net," won the Loebner Prize's Turing test competition for the third time since 2013. While the Turing test has long served as a milestone for artificial intelligence developers, advances like self-driving cars, speech processing, and image recognition have rendered the test less relevant, begging the question: What's the next moonshot goal for AI developers?

The Turing test, of course, is named for Alan Turing, who led Britain's efforts at Bletchley Park to break the Nazi's Enigma code during World War II. (Turing was played by Benedict Cumberbatch in the 2014 film, *The Imitation Game*.) In 1950, Turing asked "Can machines think?" and described a test where a computer communicating in text might display intelligent behavior like that of a human. He expected that within 50 years, such communications with a computer would be either indistinguishable from or the equivalent of those by a human.

The Turing test went on to excite the imagination of the tech community, paving the way for early chatbots like ELIZA and SmarterChild. But as AI and machine learning advance, the challenge of a machine imitating a human has become easier, and relatively trivial, compared to a machine exhibiting smarts and an ability to learn.

Steve Worswick, developer of the Mitsuku bot points to "programs like AlphaGo and the recent Dota bot that can

> **FAST FACT**
> The Loebner Prize is a spin-off of the Turing test. It offers $100,000 to the creator of an AI program that fools judges into thinking it is a human.

Alan Turing (pictured) was an English mathematician and computer scientist who created the Turing test for machine intelligence in 1950.

defeat world champions at their own field of expertise, [which] show that machines don't have to be humanlike in order to be useful."

"I believe that the Turing test goal of trying to achieve a human level of intelligence was a noble goal in its day, but computers are capable of doing so much more than a human, especially with memory and information retrieval," adds Worswick.

Beerud Sheth, CEO of Gupshup, a provider of bot building tools, says, "Turing proposed his test as a practical, simple, measurable way to evaluate machine intelligence, knowing fully well its limitations. The test is useful more as an inspirational idea than its literal interpretation as a human-imitation game."

Even so, devising software that can pass a Turing test is no small feat. For instance, Loebner prize competitors faced 20 questions ranging from current events like #8: "What do you think of Trump?" to ones requiring an understanding of context like #17: "I was trying to open the lock with the key, but someone had filled the keyhole with chewing gum, and I couldn't get it out. What couldn't I get out?"

Possible alternatives to the standard Turing test abound. For instance, Apple co-founder Steve Wozniak has suggested a coffee test, whereby a robot would be challenged to enter your home, find the kitchen and brew a cup of coffee.

"The Turing Test isn't a bad test, but it doesn't really measure intelligence," explains Ben Parr, co-founder and CMO of Octane AI, a maker of Messenger chatbots used by the likes of Maroon 5, Aerosmith, and Lindsay Lohan. (Disclosure: Octane AI is a member of the All Turtles AI startup studio.) "Clearer tests for sentience and self-awareness are needed. It may be decades or longer until we have a truly sentient machine."

Meanwhile, Kai-Fu Lee, co-founder of Sinovation Ventures and former head of Google China, argues that the Turing test simply needs updating. "I think we should stay within Turing's goal," he says, and rather than test chatbots which communicate with text, "there should be a cyborg with human skin, human vision, human speech, and human language. The test should judge the humanness or naturalness of the cyborg with all the above skills. One could add the naturalness of the skin, hair, eyes, eye-movements, body language, and more."

EVALUATING THE AUTHOR'S ARGUMENTS:

In this viewpoint Blaise Zerega writes about Alan Turing and his intelligence test for AI systems. Which expert do you think provides the most compelling response or alternative to the Turing test? Why do you think this? Do you think it's possible to create a test of artificial intelligence that won't become outdated at some point? Explain your reasoning.

Chapter 2

Are AI and Robots Beneficial to Humans?

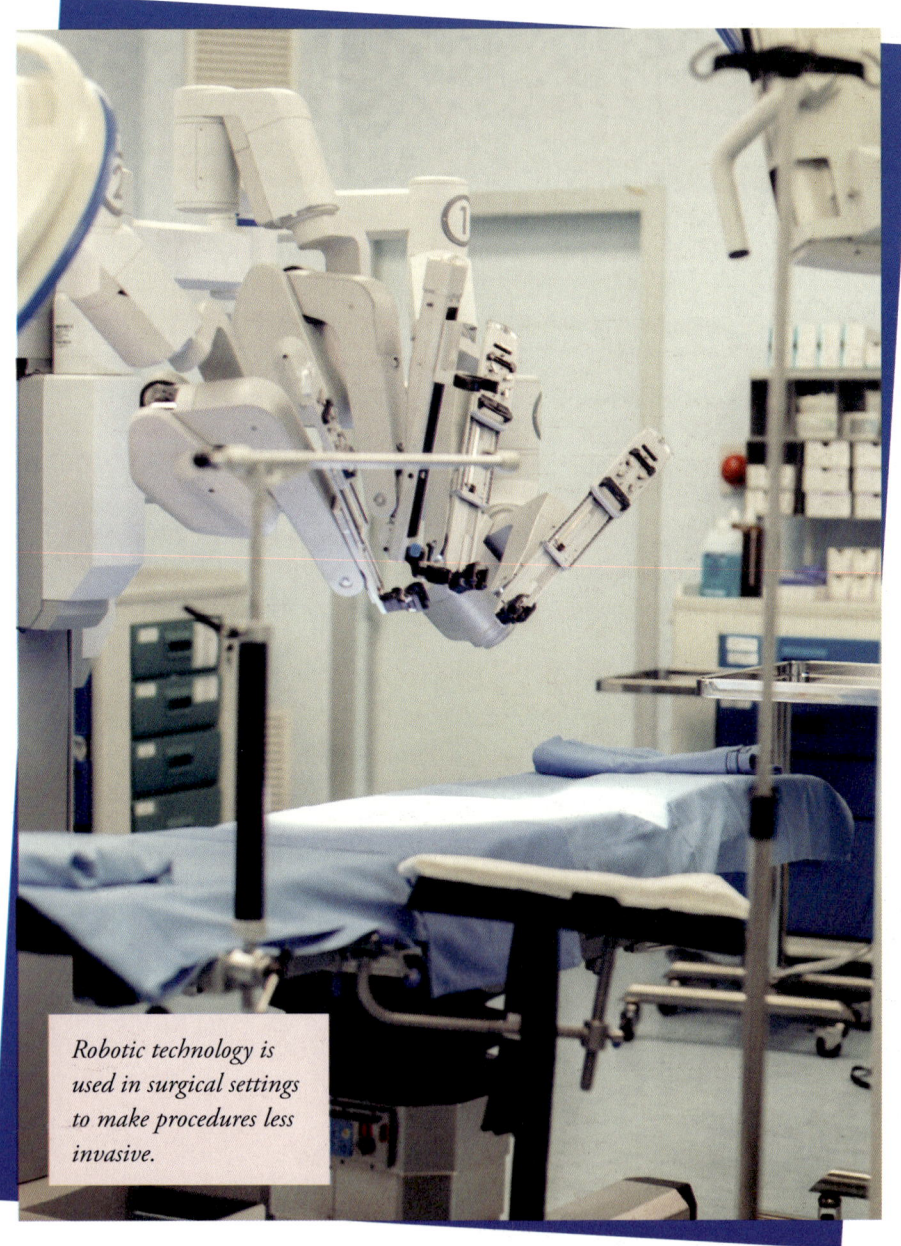

Robotic technology is used in surgical settings to make procedures less invasive.

Viewpoint 1

Can We Maintain Control Over AI?

Vivian Michaels

"Computers are processing information with unprecedented speed and accuracy, and many believe that most of our devices are outsmarting us."

In the following viewpoint, Vivian Michaels discusses the claims made by Sam Harris—a neuroscientist and philosopher—about the potential dangers AI could pose to humankind. While many fictional depictions of "evil robots" have fixated on AI with malicious intent, Harris asserts that the danger would arise from a divergence between the goals of humans and those of superintelligent machines. Michaels compares Harris' perspective to that of Stephen Hawking, who asserts that the danger of AI would come from it falling into the wrong hands and being used for destructive purposes. Vivian Michaels is a journalist who writes about technology for the *Sociable*. He is also a fitness coach.

AS YOU READ, CONSIDER THE FOLLOWING QUESTIONS:
1. What does IVA stand for?
2. Does Harris consider it useful to know how much time we have before some form of superintelligence is created?
3. What is the example the author provides of one of the worst possible applications of technological advancement?

"Is it Possible to Build AI and Not Lose Control Over It?" by Vivian Michaels, The Sociable, January 21, 2017. Reprinted with permission.

We're already seeing glimpses of what artificial intelligence (AI) is supposed to look like in our high-tech world, but can we control it?

Computers are processing information with unprecedented speed and accuracy, and many believe that most of our favorite devices are outsmarting us. And although some may see AI as something that could help us create a better world, others like Sam Harris see AI as a potential threat to humanity. In a TED Talk published last September, Sam Harris, neuroscientist and philosopher, claims that we've failed to grasp the dangers that creating AI poses to humanity.

The Slow But Sure Development of Improved AI

Artificial Intelligence is a term denoting intelligence that is exhibited by machines or that mimics humans. We're already witnesses to what is considered AI in our computers that are becoming more efficient in generating large amounts of information in the shortest span of time. But, in the near or far future, humans may develop machines that are smarter than we are, and these machines may continue to improve themselves on their own said Sam Harris in his speech on TED Talk.

The Dangers AI Poses to Humankind

Although Harris's statements sound more like science fiction than scientific observation, Harris believes that such scenarios are quite possible. The neuroscientist and philosopher explains in his TED Talk speech, that it's not that malicious armies of robots will attack us but that the slightest divergence between our goals and that of superintelligent machines could inevitably destroy us. To explain his stance, Harris explains his views on uncontrolled development of AI with an analogy of how humans relate to ants. As he puts it, we don't hate ants, but when their presence conflicts with our goals, we annihilate them. In the future, we could build machines, conscious or not, that could treat us the same way we treat ants.

Are These Scenarios Plausible?

To prove his point, Harris provided three assumptions that, if true, make a case scenario in which AI machines destroy humanity likely.

As the issues of Facebook bots and fake news have indicated, the development of AI algorithms can have negative applications as well as positive ones.

His first assumption is that intelligence is a product of information processing in physical systems, be they organic or inorganic. The second assumption is that humanity will continue to improve the intelligence of machines. His final assumption is that humans are nowhere near the peak of intelligence that machines exhibit. If all three assumptions are true, then Harris does not see why artificial intelligence turning against its creators couldn't be a possible outcome in the far future.

Other Viewpoints

But Harris isn't the only one speaking up about the dangers of uncontrolled AI development. Stephen Hawking has also joined in on the discussion recently at the opening of the new Leverhulme Centre for the Future of Intelligence (LCFI) at Cambridge University. However, the renowned physicist's viewpoints on the topic are a bit different.

According to Hawking, artificial intelligence poses a real threat to humanity when it falls into the wrong hands or when it is used in a way that won't benefit mankind. He gives examples of autonomous

> **FAST FACT**
> In 1953, the USS Mississippi test-fired one of the earliest computer-guided missiles.

weapons or new ways of the few to keep oppressing the many. So, it's not the machines that will annihilate us according to Hawking, but the wrong use of these machines. However, Hawking also notes that AI can also be of great benefit to humanity when used wisely.

Where Is Mankind Heading?

But to both Hawking and Harris, the problem seems to lie in what exactly is it that humanity is trying to achieve by improving AI machines? We've seen great technological advancements being used in the worst possible ways before (think nuclear weaponry), and it's natural that we learn from history and see AI intelligence being used in all the wrong ways.

Harris sees a problem in relying on artificial intelligence for all intellectual work as this could lead to an inequality of wealth and levels of unemployment like never before in human history. Even today, as we speak, Intelligent Virtual Assistant (IVA) software is performing tasks that not so long ago, only humans could do.

Time Matters

When we hear stories of AI taking over the planet, we usually think in time frames of maybe 50 to 100 years in the future or possibly even more. But Harris points out that thinking of how much time we have to create some form of superintelligence is not important, nor useful. The reality is that we have no idea how long it will take for humanity to create the conditions to make advanced AI. What we should focus on instead are the consequences of our striving towards better, faster, and more intelligent devices.

Conclusion

Technology has helped humankind in so many ways. But technological advancements have also created problems for our environment and with that, our safety on Earth. As we strive to improve

AI, the dangers of such efforts may start to reveal themselves. According to Harris, we may be witnesses to these dangers in the near future, and we should at least discuss the possibilities of this happening.

> **EVALUATING THE AUTHOR'S ARGUMENTS:**
>
> In this viewpoint, Vivian Michaels discusses Sam Harris and Stephen Hawking's theories regarding what they foresaw as the potential dangers of artificial intelligence. How do their theories differ? Whose do you find more convincing? Explain your reasoning.

Viewpoint 2

Artificial Intelligence Could Threaten Humanity

Rob Smith

"In the wrong hands, AI could be exploited by rogue states, terrorists and criminals."

In this viewpoint, Rob Smith outlines the ways that AI could seriously threaten the well being of society. Smith asserts that AI could be manipulated physically, digitally, and politically by rogue states, terrorists, and criminals, which could have a devastating impact on humanity. He offers particular examples of the forms these threats could take. However, Smith also outlines several possibilities for reducing this threat, which would rely on greater understanding and regulation of AI's capabilities. He reaches these conclusions through referring to the Malicious Use of Artificial Intelligence report, which was compiled by a wide range of AI experts. Rob Smith is a writer at *Formative Content*.

"3 ways AI could threaten our world, and what we need to do to stay safe," by Rob Smith, World Economic Forum, March 14, 2018. Reprinted with permission.

AS YOU READ, CONSIDER THE FOLLOWING QUESTIONS:
1. In what three areas could AI attacks occur, according to the viewpoint?
2. How does Smith suggest that physical attacks on society play out?
3. What is one way to reduce these threats as reported by the author?

Artificial intelligence (AI) could dramatically improve our lives, positively impacting everything from healthcare to security, governance and the economy. But almost all technologies can be used for ill as well as for good.

The Malicious Use of Artificial Intelligence report, compiled by experts from a number of institutions including the University of Cambridge and research firm OpenAI, argues that in the wrong hands, AI could be exploited by rogue states, terrorists and criminals.

The report outlines three areas—physical, digital and political—where AI is most likely to be exploited, and describes scenarios of how AI attacks might play out.

Remote-Controlled Car Crashes

The biggest concern involves AI being used to carry out physical attacks on humans, such as hacking into self-driving cars to cause major collisions.

"If multiple robots are controlled by a single AI system run on a centralized server, or if multiple robots are controlled by identical AI systems and presented with the same stimuli, then a single attack could also produce simultaneous failures on an otherwise implausible scale," the report states.

University of Texas Austin Professor Dr. Peter Stone, who is part of a team that recently developed a new algorithm for improving the way robots and humans communicate, thinks the report's warnings should be taken seriously, but that the situation isn't new or unique to autonomous vehicles.

While it is understandable to worry about the threats posed by the development of AI and robots, many uses of this technology are beneficial. Many robots are used to perform tasks more efficiently and accurately than humans, such as the agricultural robot pictured above.

"If someone today were to change all traffic signals in a city to be simultaneously green, disaster would ensue," Dr. Stone tells us. "And the fact that our electricity grid is fairly centralized makes us vulnerable to large-scale blackouts." According to Dr. Stone, the proper response would be stronger security measures, as well as redundancy—the provision of backup capacity—and decentralization of decision-making.

Sophisticated Phishing

In the future, attempts to access sensitive and personal information from an individual could be carried out by AI almost entirely.

"These attacks may use AI systems to complete certain tasks more successfully than any human could," the report says, adding that fraud or identity theft could become more refined and effective as AI evolves.

If most of the research and message generation typical of a phishing scam could be handled by AI, more people would be duped by this activity.

AI could impersonate people's real contacts, using a writing style that mimics the style of those contacts, making it harder to spot the scam.

Professor of Robotics at Carnegie Mellon University Illah Nourbakhsh says that since AI is rapidly evolving, we need more rapid responses to deal with its risks. "The real challenge is considering policy moves to maximize the good and minimize the bad," Nourbakhsh says. "Just as human scam artists find ever more sophisticated and nuanced ways to trick people out of their money using online scams, so AI-powered malicious actors will continuously find new pathways into our data and into our pocketbooks."

Manipulating Public Opinion

Fake news and fake videos generated by bots and AI could have a big impact on public opinion, disrupting all layers of society, from politics to media. The use of social media bots spreading fake news was already a reality during the 2016 US presidential campaign.

Well-trained bots could create a strategic advantage for political parties, and almost work as artificially intelligent propaganda machines that thrive in low-trust societies, the report claims.

This goes further than just the spread of fake text content. "AI systems can now produce synthetic images that are nearly indistinguishable from photographs, whereas only a few years ago the images they produced were crude and obviously unrealistic," the report says.

Wendell Wallach, Chair of the World Economic Forum Global Future Council on Technology, Values and Policy, and author of *A Dangerous Master: How to Keep Technology from Slipping Beyond Our Control*, says that social media is already now combining insights into human psychology and how to manipulate opinions, and it will become more sophisticated over the coming years.

"These tools will not only be used as propaganda by states to confuse and destabilize competing powers, but also as new methods employed by political leaders and political parties for tracking, manipulating and managing citizens within a country."

> **FAST FACT**
> According to a 2017 Pew Research Center survey, 56 percent of Americans do not trust self-driving cars.

Mitigating the Risks

In response to these threats, and the myriad others outlined within the report, the experts have outlined four "high-level" recommendations:

1. Policy-makers and technical researchers need to work together now to understand and prepare for the malicious use of AI.
2. Whilst AI has many positive applications, it's a dual-use technology and researchers and engineers should be mindful of and proactive about the potential for its misuse.
3. Best practices can and should be learned from disciplines with a longer history of handling dual-use risks, such as computer security.
4. The range of stakeholders engaging with preventing and mitigating the risks of malicious use of AI should be actively expanded.

"A new form of agile and comprehensive governance will be required both internationally and nationally to maximize the benefits of AI, mitigate the risks, and fulfill these four high-level recommendations," says Wendell Wallach.

At the World Economic Forum's Center for the Fourth Industrial Revolution, Head of Artificial Intelligence and Machine Learning Kay Firth-Butterfield is working on addressing some of the steps outlined in the report to mitigate the risks: "We have co-designed a project to help researchers and engineers be mindful of the misuse of AI by ensuring that teaching on culturally relevant ethical design of AI is available to any student and post-grad designing, developing and creating AI," Firth-Butterfield says.

"We're also working with governments and Boards of Directors to create best practices for the commissioning and use of AI," she explains. "We want to imagine a new type of regulator which can address the risks in an agile way, to promote and encourage use of the technology that's beneficial for the whole of humanity and the planet."

EVALUATING THE AUTHOR'S ARGUMENTS:

Rob Smith describes several ways that AI could potentially pose a threat to society. How else could AI threaten the well being of society besides the examples given in this viewpoint? What could potentially be done to reduce these risks?

Viewpoint 3

Artificial Intelligence Will Benefit Society

Kashyap Vyas

"AI is a boon to humanity and not a curse that might harm it in the future!"

In this viewpoint, Kashyap Vyas enthusiastically advocates for the continued development of AI. Vyas outlines several ways that AI is already improving human lives, and he argues that this will only continue into the future. He also discounts the cautious attitude of some, asserting that through automating processes with AI we could eliminate the possibility for human error. Furthermore, he asserts that AI development depends on both technology and creativity, which ensures that AI develops in a way that benefits and appeals to humanity. Kashyap Vyas is a thermal engineer, entrepreneur, and writer.

AS YOU READ, CONSIDER THE FOLLOWING QUESTIONS:
1. How does Vyas respond to the argument that AI will lead to the end of humanity?
2. How will weather forecasting change through the use of AI, as stated by the author?
3. According to the viewpoint, how will AI be useful in disaster relief?

"7 Ways AI Will Help Humanity, Not Harm It," by Kashyap Vyas, Interesting Engineering, December 3, 2018. Reprinted with permission.

Some robots are currently being used for emergency response. This robot is used to remove individuals from automobiles after accidents occur.

Artificial Intelligence (AI) is an intriguing concept that has fascinated experts and laymen alike for years now.

Technology in 2018 is moving at a breakneck speed, and it is safe to say that man today has significantly more power in his pocket than he had in his entire home back in the 90s.

There have been immense breakthroughs in the field of machine learning and deep learning. These concepts have allowed machines to process and analyze information themselves in a very sophisticated manner.

Thanks to these AI developments; machines can now perform complex functions such as facial recognition.

That said, there has been significant debate regarding the risks posed by Artificial Intelligence on humanity. There have been concerns about AI taking control of our lives to the extent that it proves to be detrimental to humanity.

It is also feared that as a result of the application of AI in our daily lives, unforeseen consequences can occur such as killer robots and partial election outcomes.

Although the implications of incorporating AI into our lives might sound daunting enough to eliminate its applications altogether, here is why AI is a boon to humanity and not a curse that might harm it in the future!

Enhanced Automation

Today, AI can perform intensive human labor and backbreaking tasks easily without the need for human intervention. This has immensely automated several applications and tasks in industries as well as in different sectors.

Machine learning, deep learning as well as other AI technologies are being increasingly adopted and incorporated in industries and organizations to reduce the workload of humans.

This has reduced operational costs and the cost of manpower substantially, bringing about an AI automation to a level that has not been witnessed before.

A beautiful example of the wonders of AI in enhancing the level of automation can be seen in the Japanese machine tool builder, Okuma. They recently offered a multitude of innovations to showcase the future of smart manufacturing.

This includes robots for plants of all sizes, new and improved machine tools, and smart machine tools. This clearly demonstrates the blessing of AI in the automation of industries.

Eliminates the Necessity for Humans to Perform Tedious Tasks

Artificial Intelligence can also be considered a boon to humanity given the fact that it liberates humans and allows them to perform tasks in which they excel.

We can base the necessity of AI and its applications on the argument that this technology takes care of all the tedious tasks that human must perform in order to achieve varied results.

Machines excel in taking care of cumbersome work, and this leaves enough room and time for humans to work on more creative and interpersonal aspects of their life.

Let's take the example of the banking sector that has and will be seeing a major breakthrough, thanks to the applications of AI. Financial institutions today are seen taking full advantage of this technology to make banking quicker and infinitely easier for the consumers.

This has gone a long way in helping financial analysts get some reprieve from the tedious nature of their jobs and focus on deeper research and analysis of all-around consumer experience.

Smart Weather Forecasting

In the recent few years, we have seen the use of Artificial Intelligence and its technologies in weather and climate forecasting. The field of "Climate Informatics" is constantly blossoming as it inspires a fruitful collaboration between data scientists and climate scientists.

This collaboration has come up with tools to observe and analyze increasingly complex climate data. This has helped significantly in bridging the gap between understanding and data.

There are countless applications of AI aimed for accurate weather forecasting. IBM, for instance, used its computers to improve their forecasts back in 1996.

This American multinational has ever since been refining and enhancing its forecasting methods with the incorporation of AI.

Humans now have an increased understanding of the effects and reasons for climate change.

The field of weather forecasting is super demanding and calls for intensive computing and deep-learning networks that can empower computers to dish out complex calculations.

Therefore, the advances in AI and its enviable computational power have led to the emergence of supercomputers.

This has given the man a much-needed insight into extreme climatic events so that possible disasters and natural hazards can be given a wide berth.

Next Generation Disaster Response

California saw major destruction in 2017 due to the onslaught of wildfires. More than 1 million acres worth of land was reported burned in wildfires that claimed the lives of 85 people with 249 people listed as missing.

> **FAST FACT**
>
> In a 2018 survey by the Pew Research Center, 68 percent of those surveyed said that they believe AI will improve most people's lives by 2030.

Due to the threats of climate change, more and more companies are now embracing artificial intelligence to fight disasters with algorithms.

Hence, AI has aptly demonstrated its indispensability in analyzing smart disaster responses and providing real-time data of disasters and weather events.

This is extremely useful for humans as they can detect the vulnerabilities of an area and hence help in improving disaster preparation.

The AI techniques are also beneficial as they warn us in a timely manner with enough room to organize ourselves in the face of an impending disaster and minimize loss.

It is also expected that deep-learning will soon be integrated with disaster simulations to come up with useful response strategies.

Frees Humans of The Obligation of Taking Up All Responsibilities

It is a common belief that AI will one day be the end of humanity and robots and machines will take over the planet completely and permanently.

However, what is usually ignored is the fact that the incorporation of AI in our daily lives helps free us of all responsibilities that we do not want or need.

Needless to say, we cannot let superior intelligence blindly control us. However, not employing its benefits for our advantage would be an equally ignorant thing to do.

A persuasive example in this regard is the future of war and weaponry. AI shows immense promise as a potential application in war as stipulated in Paul Scharre's book *Army of None*.

As stated in his book, in the future, militaries and machine intelligence are expected to work in tandem to conduct wars.

The Perfect Marriage of Creativity and Technology

AI can most definitely be called the perfect marriage of creativity and technology. Artificial Intelligence is nothing but a robotic machine that has the ability to think intelligently and creatively as well as translate these thoughts autonomously in varied human applications.

This fundamental basis of Artificial Intelligence is what has and can revolutionize the face of mankind. AI is not merely a unidimensional technology.

Its benefits and applications are far more important and noteworthy than its apprehensions, and this is precisely what will help humans in the future as well.

The Generation Z engagement strategies show just how AI marries creativity and technology to create the perfect results. Using the powerful tool of Artificial Intelligence, brands can now apply the correct technology to align with the needs and wishes of Gen Z.

This marketing solution that is largely data-driven is just one in a huge list of applications of AI in bringing its technological goodness and creativity together.

Zero Scope for Errors

Clive Swan, the Senior Vice President of Oracle Adaptive Intelligent Apps, shares the promise of AI and its automation as it eliminates the compulsion of human intervention and hence, removes all scope for human error.

The best thing about AI and its variety of technologies is that it is error-free. Industries and organizations usually have to leave significant room for human error because it is natural to see its presence in manual human labor.

This is what industries primarily have to deal with, and it also puts hurdles in the road to innovations as well as scientific and technological advancements.

Therefore, it is high time that we admit that we need robots and machines to provide us with a high level of precision and accuracy, leaving zero room for error.

So, there you go! These are the top 7 and super persuasive reasons why Artificial Intelligence will help humanity and not harm it in the long run.

We already see the benefits of AI in our lives in leaps and bounds, and these advantages are only likely to see fruition in the future too.

> **EVALUATING THE AUTHOR'S ARGUMENTS:**
>
> The author asserts that automation and AI would improve various processes by removing the human element, which he believes would reduce the possibility of human error. Do you agree with this argument? Why or why not? What are some examples of ways in which technology has succeeded or failed at removing the possibility for error?

Viewpoint 4

People Fear Artificial Intelligence

Dave Dickinson

Dave Dickinson assesses the fear people have toward artificial intelligence and reasons behind it. He reports on a phenomenon called the "Uncanny Valley" and its impact on how people feel about robots that look nearly human. He suggests that this discomfort toward interacting with machines has long been a pattern in human behavior, and that the negative cultural depictions of AI and robots has not done much to ease people's anxieties. However, despite these negative feelings, Dickinson notes that human behavior is often surprising, and that a significant portion of the population may appreciate and prefer these technological developments. Dave Dickinson is a science educator who has written for the science and technology section of Listosaur.com.

"Face it: most of us aren't early adopters. For whatever reason, we like to stick with the familiar and what works."

AS YOU READ, CONSIDER THE FOLLOWING QUESTIONS:
1. What does Dickinson assert is the ultimate standard for artificial intelligence?
2. How has the media portrayed AI, as reported in this viewpoint?
3. What are two reasons presented in the viewpoint for why people fear AI?

"5 Reasons People Fear Artificial Intelligence," by Dave Dickinson, Listosaur.com, November 1, 2014. Reprinted with permission.

PayPal founder and SpaceX CEO Elon Musk set the media abuzz recently when he called the development of artificial intelligence perhaps "our biggest existential threat. With artificial intelligence, we are summoning the demon." Musk further claimed that artificial intelligence, also known as AI, could be "more dangerous than nukes." Shocking stuff coming from one of the leading technology visionaries of our time. Long a staple of apocalyptic science fiction, the AI future is drawing ever closer as robots now vacuum our floors, replace factory workers and even fight our wars. But the ultimate standard in artificial intelligence, computers and robots that cannot be distinguished from humans, is still said to be at least two decades away. When that day comes, how will we react to having true artificial intelligence in the world? Here are just a few reasons that both billionaire visionaries like Musk and ordinary folks alike fear the rise of machine intelligence.

Fiction Has Not Been Kind to Robots and AI

Robots and AI computers definitely have an image problem when it comes to depictions in the media. From *Terminator* to *Alien* to *2001: A Space Odyssey*, things almost always take a turn for the worst when it comes to human-robot interactions. Even from the earliest depiction of robots in science fiction by Czech writer Karel Čapek in his 1920 play *R.U.R.* (Rossum's Universal Robots), the message has been consistent: robots and AI will compete for dominance with humans. Science fiction writer Isaac Asimov proposed his famous Three Laws of Robotics as a programming safeguard to a possible robot rebellion. (Rule No. 1: A robot may not injure a human being.)

But will machines expect equal rights, or be programmed to be happy with servitude? Is mechanical life an inevitable step after biological? Perhaps these are the questions science fiction should be asking today, to prepare us for the AI future.

People Dislike Interacting with Machines

In 1950, Alan Turing established the benchmark for AI with a concept that became known as the Turing Test. Simply put, it states that if a human interacting with a machine in a conversation could not

Fictional depictions of robots have not done much to improve how people feel about them. For instance, The Terminator *(1984) focuses on a killer robot.*

distinguish it from a person, we'd have to admit that the machine was actually thinking and capable of true intelligence. In early

> **FAST FACT**
> In 2016, Stephen Hawking said that he is almost certain a major technological disaster will threaten humanity in the next 1,000 to 10,000 years.

interactions between machines and humans, there was no question we were dealing with a machine (remember those talking Coke machines from the 1980s?) Sure, we now talk to our mobile phones via Siri and our GPS devices give us directions, but many people are still Luddites when it comes to even using the self-checkout line. Most baseball fans blanch at the very mention of balls and strikes being called by a computer; while more accurate, some critics argue such a system would be a slap in the face of tradition.

At this point, it's tough to tell how we'll integrate AI into our lives. Psychology dictates that we like to deal with a presence with a capacity for a moral compass. For example, we're likely to take it as fate if a tornado wipes out our home, but we feel slighted if someone cuts in line in front of us. It remains to be seen just what side of the line AI will ultimately fall on in most people's minds. Still, speaking computers now being used for automated surveys are getting frighteningly intelligent, and imagine what a game changer it might be online if spambots reach a point where they are indistinguishable from humans.

Many Assume Robots Will Take Away Jobs

This is an immediate fear for many people. Look no further than the automotive industry, where futuristic-looking robots assemble cars, with nary a human in sight. AI will eventually threaten other forms of employment as well. In fact, computerized news aggregators are already being used, for example, to narrate sports highlights clips; it's not tough to say "Team X beat Team Y by x number of points," over and over again. Will we one day reach a stage where an AI robot is deemed much more suitable to conduct a delicate brain or heart surgery than a human? On the plus side, perhaps robots will ultimately free us up from the kind of drudge work that most of us really don't want to do in the first place.

Fear of the Unknown

Face it: most of us aren't early adopters. For whatever reason, we like to stick with the familiar and what works. Then there are those who are already in charge with a vested interest in the status quo, and are likely to also actively discourage change. But there is a tipping point, and AI and robotics may soon experience this drive from resistance to acceptance. In the case of mobile phones and the Internet, the tipping point came when there was a perceived advantage and when it was realized that you couldn't afford not to play the game. Will people still fear robots and AI when they see others using them to get ahead?

The robotic revolution may also play a part in what's being termed the "singularity," a point at which technology may become self-sustaining and self-propagating and simply be out of our hands. When computers become sentient, what will they do with humans? Musk's recent comments were concerned specifically with the use of robotics and artificial intelligence in warfare. He's certainly not the only one with such fears. A 2009 report commissioned by the US Navy noted several possible scenarios in which AI robots could turn on humans (a la in the *Terminator* series). Huge coding projects, involving hundreds of programmers and millions of lines of code, raise the risk a malfunction could arise. Terrorists could also reprogram the machines to attack their masters. One of the authors of the report said it will be critical to program in a "code of ethics" to control the robots.

Fear of the "Uncanny Valley"

Turns out that robots that look almost but not quite human tend to creep us out. Robots that look almost human have entered an area of perception known as the Uncanny Valley, a point at which they have many human-like traits but they're just un-human enough to make us feel uncomfortable. And yes, there is an actual "fear of robots," of sorts known as automatonophobia, which encompasses the fear of anything mimicking a person (ventriloquist dummies, wax statues, etc.).

The Japanese have created several human-looking robots that are indeed eerie to watch in action. In the case of the Uncanny Valley effect, it's often unclear just what sparks these feelings of repulsion: we only know when interacting with human-like robots that something is wrong. Is it the fact that they don't react to their surroundings in the way that we expect? Or are they too perfect a recreation of us? Perhaps, most future robots won't even need to take human form, except maybe the ones that we'll need to interact with on a daily basis. Still, human nature has surprised us before, and many seem to now prefer seeking out interaction via Facebook and Twitter versus face-to-face … will a segment of the population ultimately prefer the simplistic perceived perfection of a human-robotic relationship to the often complex and messy reality of human social interplay?

EVALUATING THE AUTHOR'S ARGUMENTS:

In this viewpoint Dave Dickinson reports that many people fear AI and its consequences. Based on the reasoning he offers in the viewpoint, do you think he considers these fears to be reasonable? Can you think of other reasons someone might be afraid of artificial intelligence besides those Dickinson mentioned?

Viewpoint 5

Artificial Intelligence Can Save Our Planet

Celine Herweijer

"It is now possible to tackle some of the world's biggest problems with emerging technologies such as AI. It's time to put AI to work for the planet."

In the following viewpoint, Celine Herweijer paints a bright picture of how artificial intelligence can be used to improve life on Earth. Herweijer believes that AI can help with problems occurring around the world and can help facilitate breakthroughs in the field of Earth sciences. Herweijer backs up this claim by pointing out ways that AI can make a difference in transportation, agriculture, energy, climate, weather, disaster response, and city infrastructure. Celine Herweijer is a partner at PwC United Kingdom, a multinational professional services network, where she co-heads the climate change and international development division. She holds a PhD in computer modeling and Earth sciences from Columbia University.

AS YOU READ, CONSIDER THE FOLLOWING QUESTIONS:
1. According to the author, what are the risks associated with AI?
2. How can AI help improve air quality, as outlined by Herweijer?
3. What can AI do for climate and weather, according to this viewpoint?

"8 ways AI can help save the planet," by Celine Herweijer, World Economic Forum, January 24, 2018. Reprinted with permission.

It's a historic moment for Artificial Intelligence (AI). All the pieces are coming together: big data, advances in hardware, emerging powerful AI algorithms, and an open source community for tools that reduces barriers to entry for industry and start-ups alike. The result: AI is being propelled out of research labs and into our everyday lives, from navigating cities, ride shares, our energy networks, to the online world.

In 2018 everyone is starting to see the business value of AI. It is being added to more and more things every year, and it is getting smarter and smarter—accelerating human innovation. But as AI becomes more powerful, more autonomous and broader in its use and impact, the unsolved issue of AI safety is paramount. Risks include: bias, poor decision making, low transparency, job losses and malevolent use of AI, such as autonomous weaponry.

The challenge, however, goes beyond guiding "human friendly AI" to ensuring "Earth friendly AI." As the scale and urgency of the economic and human health impacts from our deteriorating natural environment grows, we have an opportunity to look at how AI can help transform traditional sectors and systems to address climate change, deliver food and water security, build sustainable cities, and protect biodiversity and human wellbeing.

To this end, in a new Forum-PwC report launched at Davos this year, we showcase the significant opportunity to harness AI for the Earth. Here we outline eight of the identified "game changer" AI applications to address this planet's challenges:

Autonomous and Connected Electric Vehicles

AI-guided autonomous vehicles (AVs) will enable a transition to mobility on-demand over the coming years and decades. Substantial greenhouse gas reductions for urban transport can be unlocked through route and traffic optimisation, eco-driving algorithms, programmed "platooning" of cars to traffic, and autonomous ride-sharing services. Electric AV fleets will be critical to deliver real gains.

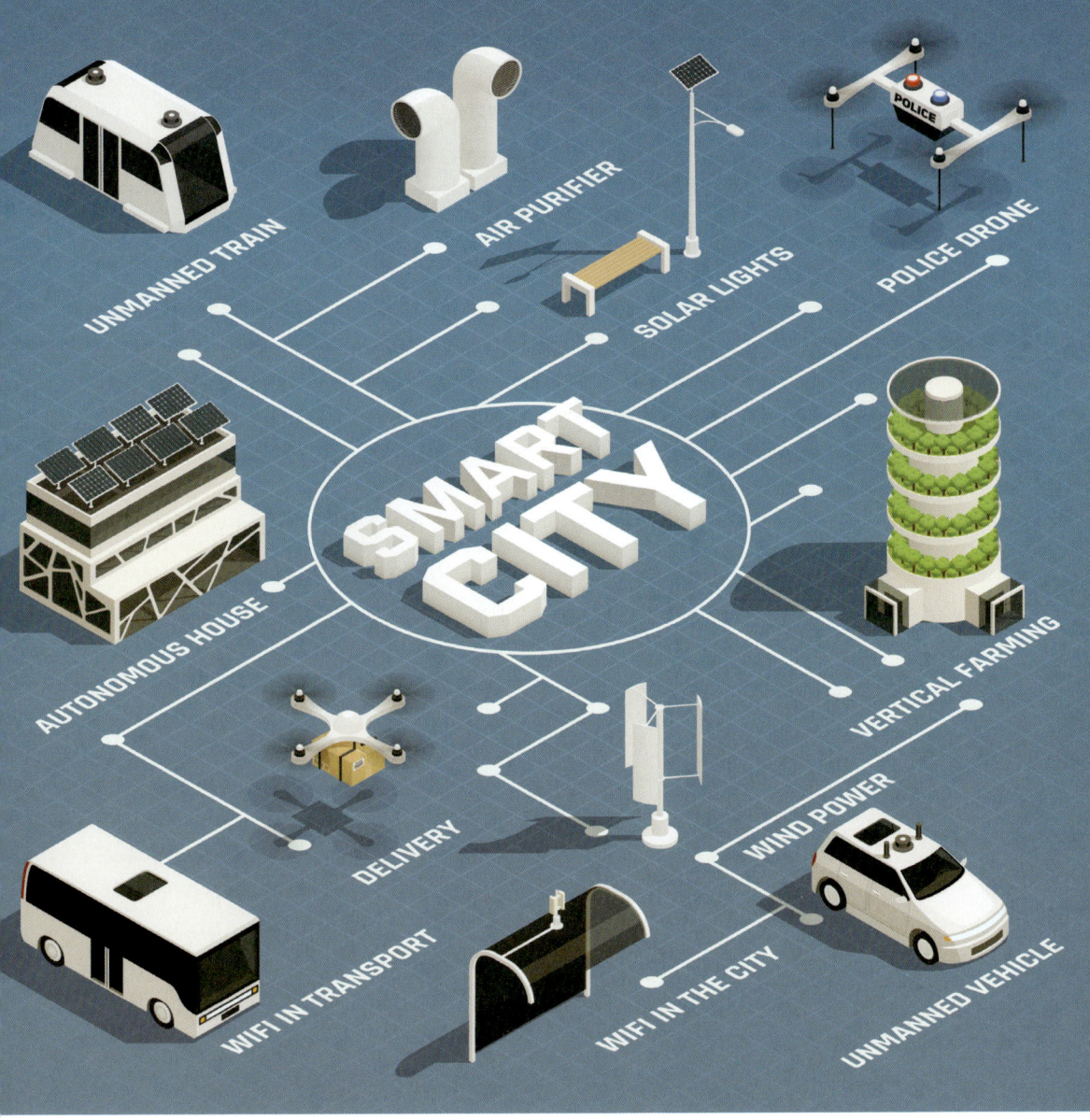

Smart cities could use technology to make various aspects of life more efficient and environmentally friendly. Some of the potential applications are pictured above.

Distributed Energy Grids

AI can enhance the predictability of demand and supply for renewables across a distributed grid, improve energy storage, efficiency and load management, assist in the integration and reliability of renewables and enable dynamic pricing and trading, creating market incentives.

> **FAST FACT**
> Several top car companies—including GM, Ford, and Audi—expect autonomous cars will be available to the public by around 2020.

Smart Agriculture and Food Systems

AI-augmented agriculture involves automated data collection, decision-making and corrective actions via robotics to allow early detection of crop diseases and issues, to provide timed nutrition to livestock, and generally to optimise agricultural inputs and returns based on supply and demand. This promises to increase the resource efficiency of the agriculture industry, lowering the use of water, fertilisers and pesticides which cause damage to important ecosystems, and increase resilience to climate extremes.

Next Generation Weather and Climate Prediction

A new field of "Climate Informatics" is blossoming that uses AI to fundamentally transform weather forecasting and improve our understanding of the effects of climate change. This field traditionally requires high performance energy-intensive computing, but deep-learning networks can allow computers to run much faster and incorporate more complexity of the "real-world" system into the calculations.

In just over a decade, computational power and advances in AI will enable home computers to have as much power as today's supercomputers, lowering the cost of research, boosting scientific productivity and accelerating discoveries. AI techniques may also help correct biases in models, extract the most relevant data to avoid data degradation, predict extreme events and be used for impacts modelling.

Smart Disaster Response

AI can analyse simulations and real-time data (including social media data) of weather events and disasters in a region to seek out vulnerabilities and enhance disaster preparation, provide early warning, and prioritise response through coordination of emergency information capabilities. Deep reinforcement learning may one day be integrated into disaster simulations to determine optimal response strategies, similar to the way AI is currently being used to identify the best move in games like AlphaGo.

AI-Designed Intelligent, Connected and Livable Cities

AI could be used to simulate and automate the generation of zoning laws, building ordinances and floodplains, combined with augmented and virtual reality (AR and VR). Real-time city-wide data on energy, water consumption and availability, traffic flows, people flows, and weather could create an "urban dashboard" to optimise urban sustainability.

A Transparent Digital Earth

A real-time, open API, AI-infused, digital geospatial dashboard for the planet would enable the monitoring, modelling and management of environmental systems at a scale and speed never before possible—from tackling illegal deforestation, water extraction, fishing and poaching, to air pollution, natural disaster response and smart agriculture.

Reinforcement Learning for Earth Sciences Breakthroughs

This nascent AI technique—which requires no input data, substantially less computing power, and in which the evolutionary-like AI learns from itself—could soon evolve to enable its application to real-world problems in the natural sciences. Collaboration with Earth scientists to identify the systems—from climate science, materials science, biology, and other areas—which can be codified to apply reinforcement learning for scientific progress and discovery is vital. For example, DeepMind co-founder, Demis Hassabis, has suggested that in materials science, a descendant of AlphaGo Zero could be used to search for a room temperature superconductor—a hypothetical substance that allows for incredibly efficient energy systems.

To conclude, we live in exciting times. It is now possible to tackle some of the world's biggest problems with emerging technologies such as AI. It's time to put AI to work for the planet.

EVALUATING THE AUTHOR'S ARGUMENTS:

Celine Herweijer asserts that the focus in AI development has been on creating human-friendly AI rather than climate-friendly AI, and that priorities should shift to focus more on the potential environmental benefits of AI. Do you agree with the author's claim? If you agree with the author, why do you think the focus has been primarily on how AI will impact humans?

Chapter 3

How Should AI and Robots Be Regulated to Benefit the Future of the Human Race?

Drones are restricted from flying in certain areas, showing how robots are already being regulated.

Viewpoint 1

Should Artificial Intelligence Be Regulated?

Greg Benson

In the following viewpoint, Greg Benson asks a simple question: should AI be regulated? Benson then analyzes several scenarios surrounding the concept of AI and its regulation, providing answers to some of the main questions and objections that have been raised regarding the future of AI. Some interesting questions brought up by Benson include whether AI could have the ability to commit crimes, as well as whether it could be capable of judging and governing. Because of how quickly technology has developed and evolved, Benson asserts that the government has not been able to keep up in terms of regulation, though he thinks that greater efforts must be made by everyone to keep up with AI developments. Dr. Greg Benson is a professor of computer science at the University of San Francisco.

"Can and should there be regulators to control AI?"

"Governing AI: Can Regulators Control Artificial Intelligence?" by Greg Benson, Professor of Computer Science, University of San Francisco Chief Scientist, SnapLogic Inc., June 7, 2018. Reprinted with permission.

AS YOU READ, CONSIDER THE FOLLOWING QUESTIONS:
1. What kind of AI do Siri and Alexa use, according to the viewpoint?
2. According to Benson, is a self-driving car that "decides" to brake expressing independent thought?
3. What role should education play, according to the author?

Artificial Intelligence technology is growing rapidly, and with that growth comes the fear that it will get out of hand. Can and should there be regulators to control AI?

How do you see the world adapting/evolving in an AI environment?

In terms of computer applications, we will see increasing application and adoption of machine learning (ML) and artificial intelligence (AI) techniques.

We already see this in shopping recommendations, games, and large social networks. Voice assistants such as Siri, Alexa, and Google Assistant use ML to perform natural language processing and classification to respond appropriately.

Such techniques will make interacting with devices increasingly seamless, which will ultimately make technology easier to use while making humans more efficient in finding and managing information.

Just like automation and the application of machine learning to businesses and business processes, there are many opportunities to apply similar techniques to government functions. Transportation, public utilities, and public health are all massive public-sector functions that could benefit from the application of ML and AI.

Stephen Hawking says there could be a robot apocalypse. What are risks associated with developing software capable of decision-making and "independent" thought?

I've been careful to use the term machine learning over artificial intelligence because we have not yet achieved what could be considered "independent" thought in computer software. Also,

Voice assistants like the Amazon Alexa are examples of how AI is used to make everyday life easier.

decision-making and "independent" thought are two different concepts.

We now have computer-enhanced cars that can "decide" to apply the brakes to avoid a collision (automatic emergency braking), but such a "decision" is limited to a very specific situation. In my view, independent thought is associated with self-awareness and emotion. We have not achieved this type of AI to date, and it seems we are a long way off.

We don't know yet if it is even possible. I'm skeptical that we will be able to develop computers that achieve true independent thought. The complex interactions of our brain functions with our physiology seem truly difficult to replicate.

That said, using computers to automate complex physical systems, such as self-driving cars, that require "judgment" will be tricky. For example, a self-driving car could have to decide between potentially harming the passengers or perhaps many more pedestrians.

Should the car protect the passengers at all costs or try to minimize the total harm to all the humans involved even if that means harming the passengers? If you knew that a self-driving car was programmed to minimize the total harm to human life in certain situations, would you agree to allow the car to drive you?

So there are definitely risks in allowing software to control physical systems. I think our adoption of computer automation will be evaluated on a case-by-case basis. In some cases, increased automation will make human activities safer. In other cases, we may choose to continue relying on human decision making but perhaps augmented with computer assistance.

Will AI be capable of governing (in civil disputes, for example)?

This gets back to the notion of judgment and, in addition, morality. If true AI is possible and it could assess situations and pass judgments, then we would evaluate the AI for governing just as we would human judges.

Just like human judges must convince the public regarding their suitability for upholding the law, so would AI judges. If they were to pass our current tests for appointing judges, then it may be acceptable to allow an AI to govern. That said, will AI be capable of such governing? I don't think so.

AI has been used in recent elections (US/UK—to gauge popular policies and influence voters); will this be a theme in the future?

To be precise, I don't think true AI was used in these situations, but rather the narrower application of machine learning. Humans controlled the execution of machine learning algorithms. AI did not choose by itself to influence elections. I believe machine learning will be increasingly used as a tool by humans to influence voters and to shape policy.

Do you think governance will need to adapt to handle AI? Is it even possible to regulate it?

AI, in practice, is really the application of algorithms to data in a process that is controlled by humans. So, in this sense, governance needs to adapt to handle and regulate computer software that is used in activities that can impact human well-being such as voting machines, transportation, health systems, and many others.

Computer technology has advanced at such a rapid pace, government oversight has not been able to keep up. It is interesting to think that to build a bridge you must be a licensed mechanical engineer, however, software developers require no such license to work on many types of systems that can affect human life, such as medical devices.

Can we have governance for computer software without stifling innovation and delaying potential benefits to human life? I'm not sure.

Do you think governments will have a say in what technology is developed legally?

I think so indirectly in the sense that results or impact of technology will be allowed through the law, or not. Consider the murky area of drones. Governments will certainly have a say on how drone technology can be deployed and utilized.

The citizens will demand regulation through the democratic process. Although in the case of drones, the technology is so new and different, our understanding of their impact and laws to regulate them have not yet caught up, so this will take time.

Will robots ever be capable of committing a crime?

If you believe we can create conscious artificial intelligent robots that can have flaws just like humans, then yes, such robots could commit crimes just like humans. This would suggest that robots have emotion and selfish motives. I'm doubtful we will achieve such consciousness in digital technology.

Perhaps subtler is whether or not autonomous systems supported by AI and ML can commit a crime, but not knowingly. I think this is a more likely scenario. For example, a drone may fly into airspace or property where no trespassing is allowed. An autonomous car may exceed the speed limit accidentally or because it was necessary to avoid an accident.

> **FAST FACT**
>
> In 2017, only 27 percent of Americans surveyed by Pew Research said that robots have positively affected their work.

In your view, are governments on the back foot when it comes to regulating AI?

Software, in general, is evolving at such a rapid pace, and this will continue to present ongoing challenges to government regulation.

For example, software can be developed to exhibit bias, lead to unsafe systems, or make financially irresponsible decisions. So, the government needs to stay close as all technology develops and step in accordingly to regulate where it makes sense. AI is the most recent and prominent example.

What do they need to do, to catch up with all the developments?

I think we need to do a better job of educating the larger population on these complex technology topics. We need more people with a greater understanding of technology, software, and mathematics. In fact, we need to develop technology that can help us better understand the risks and benefits of existing and future software systems.

In an AI future, will a universal income be needed? If so, what might it look like, and how could it be administered?

I believe that AI technology will augment human activity more than replace it. It will make us more efficient so that we can use our judgment in different ways.

Such augmentation will shift how human intelligence is applied and human creativity is realized. Such new forms of human-computer interaction will result in income earning activities. I don't believe a universal income will be needed.

> **EVALUATING THE AUTHOR'S ARGUMENTS:**
>
> In this viewpoint Greg Benson suggests that artificial intelligence will augment human activities—such as jobs—instead of merely replacing them. What do you think? Do you agree with Benson? Why or why not?

Will AI Make a World Without Work Possible?

Ryan Avent

"Despite impressive progress in robotics and machine intelligence, those of us alive today can expect to keep on labouring until retirement."

In this viewpoint, Ryan Avent analyzes a question that is of concern to many forward-thinking individuals: what will become of society as artificial intelligence displaces workers? Avent maintains that other sources will have to be found for the things that people receive through their jobs, especially their wages. Avent argues that the rich will initially become richer and that many people will have to rely on social systems for their means of living. However, he asserts that we still have some time to work this out politically. Ryan Avent is a senior editor and columnist at the *Economist*. His is also the author of *The Wealth of Humans: Work and Its Absence in the Twenty-First Century*.

AS YOU READ, CONSIDER THE FOLLOWING QUESTIONS:

1. According to Avent, what do many workers receive from their jobs besides money?
2. What are the effects of the digital revolution presented in this viewpoint?
3. How long does Avent predict it will take society to come up with positive solutions to this dilemma?

"A world without work is coming – it could be utopia or it could be hell," by Ryan Avent, Guardian News and Media Limited, September 19, 2016. Reprinted with permission.

With the advancement of robotic and AI technology, it is becoming increasingly likely that jobs will no longer have to be performed by humans, which would result in more leisure time.

Most of us have wondered what we might do if we didn't need to work—if we woke up one morning to discover we had won the lottery, say. We entertain ourselves with visions of multiple homes, trips around the world or the players we would sign after buying Arsenal. For many of us, the most tantalising aspect of such visions is the freedom it would bring: to do what one wants, when one wants and how one wants.

But imagine how that vision might change if such freedom were extended to everyone. Some day, probably not in our lifetimes but perhaps not long after, machines will be able to do most of the tasks that people can. At that point, a truly workless world should be possible. If everyone, not just the rich, had robots at their beck and call, then such powerful technology would free them from the need to submit to the realities of the market to put food on the table.

Of course, we then have to figure out what to do not only with ourselves but with one another. Just as a lottery cheque does not free the winner from the shackles of the human condition, all-purpose machine intelligence will not magically allow us all to get along. And what is especially tricky about a world without work is that we must begin building the social institutions to survive it long before the technological obsolescence of human workers actually arrives.

Despite impressive progress in robotics and machine intelligence, those of us alive today can expect to keep on labouring until retirement. But while Star Trek-style replicators and robot nannies remain generations away, the digital revolution is nonetheless beginning to wreak havoc. Economists and politicians have puzzled over the struggles workers have experienced in recent decades: the pitiful rate of growth in wages, rising inequality, and the growing flow of national income to profits and rents rather than pay cheques. The primary culprit is technology. The digital revolution has helped supercharge globalisation, automated routine jobs, and allowed small teams of highly skilled workers to manage tasks that once required scores of people. The result has been a glut of labour that economies have struggled to digest.

Labour markets have coped the only way they are able: workers needing jobs have little option but to accept dismally low wages. Bosses shrug and use people to do jobs that could, if necessary, be done by machines. Big retailers and delivery firms feel less pressure to turn their warehouses over to robots when there are long queues of people willing to move boxes around for low pay. Law offices put off plans to invest in sophisticated document scanning and analysis technology because legal assistants are a dime a dozen. People continue to staff checkout counters when machines would often, if not always, be just as good. Ironically, the first symptoms of a dawning era of technological abundance are to be found in the growth of low-wage, low-productivity employment. And this mess starts to reveal just how tricky the construction of a workless world will be. The most difficult challenge posed by an economic revolution is not how to come up with the magical new technologies in the first place; it is how to reshape society so that the technologies can be put to good use while also keeping the great mass of workers satisfied with their lot in life. So far, we are failing.

Preparing for a world without work means grappling with the roles work plays in society, and finding potential substitutes. First and foremost, we rely on work to distribute purchasing power: to give us the dough to buy our bread. Eventually, in our distant Star

> **FAST FACT**
> Amazon began using robots to fulfill orders in 2013. It now has around 100,000 robotic systems in twenty-six fulfillment centers.

Trek future, we might get rid of money and prices altogether, as soaring productivity allows society to provide people with all they need at near-zero cost.

For a good while longer, wages will continue to be the main way people come by money, and prices will be needed to ration access to scarce goods and services. But in the absence of any broader social change, pushing people out of work will simply redirect the flow of income from workers to firm-owners: the rich will get richer. Freeing people from work without social collapse will therefore require society to find ways other than pay for labour to channel money to those not on the job. People might come to receive more of their income in the form of state-led redistribution: through the payment of a basic income, for instance, or direct public provision of services such as education, healthcare and housing. Or, perhaps, everyone could be given a capital allotment at birth.

These sorts of arrangements don't magically materialise as machines become more powerful. They must be brought into existence through political action. And that's where things start to get complicated. One problem is that large-scale social overhaul takes a long time to emerge and have an effect. Another is that money for nothing is not necessarily what the displaced masses are interested in.

Ongoing political debates illustrate the problem. There are lots of ways a government could boost workers' pay. It could raise the minimum wage, increase wage subsidies, enact a basic income, or use more heavy-handed regulation to protect industries and force firms to share more of their profits with labourers. Tellingly, workers and trade unions seem least interested in the policies, such as a basic income, that break the link between compensation and work. This makes the building of our eventual utopia tricky; a hefty rise in the minimum wage would benefit lots of workers, but it would also discourage some firms from using the cheap labour they have been soaking up, forcing the jobless to get along in a world in which they cannot find work yet also lack the monetary means to stay out of poverty.

Workers' preferences are easy to understand. Work is not just a means for distributing purchasing power. It is also among the most important sources of identity and purpose in individuals' lives. If the role of work in society is to shrink, other sources of purpose and identity will need to grow. Some people will manage to find these things for themselves: pursuing passions too uneconomic to live on or engaging in voluntarism, just as many retirees find satisfying ways to fill their days. But others will find themselves at a loss.

Workers are sure to feel uncomfortable with reforms designed to clear a path to their own economic irrelevance. They are not the only ones likely to object. Redistribution implies taking as well as giving. And while some tech entrepreneurs seem to be warming to the idea of something like a universal basic income, perhaps seeing it as a moral licence to disrupt, the reservoir of resentment at those perceived to be getting too good a deal from the government never runs dry. Rich Americans are already annoyed enough at the "takers" among their countrymen, those Mitt Romney labelled an incorrigible 47% in his 2012 campaign for the presidency, who pay no federal income tax—even though most work, pay other taxes, and are simply too poor to owe any income tax to the federal government. The haves who will inevitably provide a disproportionate share of the funding for future welfare states will need convincing to part with their cash.

So societies might decide that people must be made to contribute in some way to the community to qualify for state support. Those not in work, for example, might have to participate in community service or other activity. Another approach might be more seductive. Those still in work might be less grumpy about funding a more generous welfare state if beneficiaries are deemed to be enough like them: fellow tribesmen, people of similar background and therefore felt to be deserving of charity.

Around the rich world, it is interesting to note that it is not so much the generosity of state redistribution that is provoking societal unrest, but the fact that out groups—from Latinos to Poles to refugees—are grabbing a share.

Building a workless utopia in which wealth is broadly shared, people are mostly satisfied with their lot in life, and the peace isn't kept by excluding any inconvenient foreigners, is no easy task. The grappling has already begun, and the initial rounds of negotiation are more than a little discouraging. Two centuries from now, I am confident, we will have worked everything out splendidly. Assuming, that is, that those of us alive now can manage the first painful steps without wrecking the world in the process.

> **EVALUATING THE AUTHOR'S ARGUMENTS:**
>
> In this viewpoint, Ryan Avent envisions a world in which most people will not work because of advances in artificial intelligence. Do you agree with his reasoning? Would you appreciate this scenario? Why or why not?

Viewpoint 3

Will Robots Replace Human Employees?

Owen Bowcott

"Jobs at all levels in society presently undertaken by humans are at risk of being reassigned to robots or AI."

In the following viewpoint, Owen Bowcott writes about a future in which many employees may be replaced by robots. Bowcott maintains that governments will need to step in and regulate this phenomenon, determining which jobs should belong strictly to people. He provides examples of situations where robots are already being used around the world in great numbers and concludes that the world of work as we know it is disintegrating. Owen Bowcott is a legal affairs correspondent at the *Guardian*. He also co-wrote *Beating the System*, a book about computer hacking.

AS YOU READ, CONSIDER THE FOLLOWING QUESTIONS:
1. According to Bowcott, how does the cost of using robots as workers compare to human employees?
2. Does the author claim that all jobs will be taken over by robots?
3. As reported in the viewpoint, which country has the highest proportion of robots to people?

"Rise of robotics will upend laws and lead to human job quotas, study says," by Owen Bowcott, Guardian News and Media Limited, April 4, 2017. Reprinted with permission.

With robots taking over an increasingly large number of jobs, hiring processes for human employees could become even more competitive.

Innovation in artificial intelligence and robotics could force governments to legislate for quotas of human workers, upend traditional working practices and pose novel dilemmas for insuring driverless cars, according to a report by the International Bar Association.

The survey, which suggests that a third of graduate level jobs around the world may eventually be replaced by machines or software, warns that legal frameworks regulating employment and safety are becoming rapidly outdated.

The competitive advantage of poorer, emerging economies—based on cheaper workforces—will soon be eroded as robot production lines and intelligent computer systems undercut the cost of human endeavour, the study suggests.

While a German car worker costs more than €40 (£34) an hour, a robot costs between only €5 and €8 per hour. "A production robot is thus cheaper than a worker in China," the report notes. Nor does a robot "become ill, have children or go on strike and [it] is not entitled to annual leave."

The 120-page report, which focuses on the legal implications of rapid technological change, has been produced by a specialist team of employment lawyers from the International Bar Association, which acts as a global forum for the legal profession.

The report covers both changes already transforming work and the future consequences of what it terms "industrial revolution 4.0." The three preceding revolutions are listed as: industrialisation, electrification and digitalisation. "Industry 4.0" involves the integration of the physical and software in production and the service sector. Amazon, Uber, Facebook, "smart factories" and 3D printing, its says, are among current pioneers.

The report's lead author, Gerlind Wisskirchen—an employment lawyer in Cologne who is vice-chair of the IBA's global employment institute, said: "What is new about the present revolution is the alacrity with which change is occurring, and the broadness of impact being brought about by AI and robotics.

"Jobs at all levels in society presently undertaken by humans are at risk of being reassigned to robots or AI, and the legislation once in place to protect the rights of human workers may be no longer fit for purpose, in some cases … New labour and employment legislation is urgently needed to keep pace with increased automation."

Peering into the future, the authors suggest that governments will have to decide what jobs should be performed exclusively by humans—for example, caring for babies. "The state could introduce a kind of 'human quota' in any sector," and decide "whether it intends to introduce a 'made by humans' label or tax the use of machines," the report says.

Increased mechanical autonomy will cause problems of how to define legal responsibility for accidents involving new technology such as driverless cars. Will it be the owner, the passengers, or manufacturers who pay the insurance?

"The liability issues may become an insurmountable obstacle to the introduction of fully automated driving," the study warns. Driverless forklifts are already being used in factories. Over the past 30 years there have been 33 employee deaths caused by robots in the US, it notes.

Limits, it says, will have to be imposed on some aspects of machine autonomy. The study adopts the military principle, endorsed by the Ministry of Defence, that there must always be a "human in

> **FAST FACT**
> As reported by Pew Research, 18 percent of Americans personally know someone who has lost a job or had their hours or pay reduced because of some kind of automation.

the loop" to prevent the development and deployment of entirely autonomous drones that could be programmed to select their own targets.

"A no-go area in the science of AI is research into intelligent weapon systems that open fire without a human decision having been made," the report states. "The consequences of malfunctions of such machines are immense, so it is all the more desirable that not only the US, but also the United Nations discusses a ban on autonomous weapon systems."

The term "artificial intelligence" (AI) was first coined by the American computer scientist John McCarthy in 1955. He believed that "every aspect of learning or any other feature of intelligence can in principle be so precisely described that a machine can be made to simulate it." Software developers are still attempting to achieve his goal.

The gap between economic reality in the self-employed "gig economy" and existing legal frameworks is already growing, the lawyers note. The new information economy is likely to result in more monopolies and a greater income gap between rich and poor because "many people will end up unemployed, whereas highly qualified, creative and ambitious professionals will increase their wealth."

Among the professions deemed most likely to disappear are accountants, court clerks and "desk officers at fiscal authorities."

Even some lawyers risk becoming unemployed. "An intelligent algorithm went through the European Court of Human Rights' decisions and found patterns in the text," the report records. "Having learned from these cases, the algorithm was able to predict the outcome of other cases with 79% accuracy … According to a study conducted by [the auditing firm] Deloitte, 100,000 jobs in the English legal sector will be automated in the next 20 years."

The pioneering nation in respect of robot density in the industrial sector is South Korea, which has 437 robots for every 10,000 employees in the processing industry, while Japan has 323 and Germany 282.

Robots may soon invade our home and leisure environments. In the "Henn-na Hotel" in Sasebo, Japan, "actroids"—robots with a human likeness—are deployed, the report says. "In addition to receiving and serving the guests, they are responsible for cleaning the rooms, carrying the luggage and, since 2016, preparing the food."

The robots are able to respond to the needs of the guests in three languages. The hotel's plan is to replace up to 90% of the employees by using robots in hotel operations with a few human employees monitoring CCTV cameras to see whether they need to intervene if problems arise.

The traditional workplace is disintegrating, with more part time employees, distance working, and the blurring of professional and private time, the report observes. It is being replaced by "the 'latte macchiato' workplace where employees or freelance workers in the cafe around the corner, working from their laptops."

The workplace may eventually only serve the purpose of maintaining social network between colleagues.

EVALUATING THE AUTHOR'S ARGUMENTS:

Owen Bowcott suggests that governmental regulations may play a role in deciding which workers will be entirely replaced by robots. Do you think Greg Benson, the author of Viewpoint 1 in Chapter 3, would agree? Use examples from both viewpoints to support your answer.

Viewpoint 4

Take Control of Autonomous Weapons Before It's Too Late

"Also known as killer robots, these AI-powered ships, tanks, planes and guns could fight the wars of the future without any human intervention."

Mattha Busby

In this viewpoint, Mattha Busby reports on a frightening futuristic phenomenon: the autonomous killer robot. According to Busby, these killer robots will have no qualms about killing because they will lack any moral conscience, and in effect will kill faster than humans will be able to defend themselves. Busby provides examples of autonomous weapons in various countries and perspectives from political leaders around the world on how they see this issue. Though efforts are being made to prepare for the destructive capabilities of autonomous weapons—as demonstrated by a meeting between nations that took place at the UN in 2018—reaching a consensus on how to address the issue has proven challenging. Mattha Busby is a freelance journalist who has had work published in the *Guardian*.

"Killer robots: pressure builds for ban as governments meet," by Mattha Busby, Guardian News and Media Limited, April 9, 2018. Reprinted with permission.

AS YOU READ, CONSIDER THE FOLLOWING QUESTIONS:
1. What does "Laws" stand for in this viewpoint?
2. Which countries are already developing AI-inspired military systems?
3. According to this viewpoint, will political realities interfere with international cooperation to regulate this type of warfare?

They will be "weapons of terror, used by terrorists and rogue states against civilian populations. Unlike human soldiers, they will follow any orders however evil," says Toby Walsh, professor of artificial intelligence at the University of New South Wales, Australia.

"These will be weapons of mass destruction. One programmer and a 3D printer can do what previously took an army of people. They will industrialise war, changing the speed and duration of how we can fight. They will be able to kill 24-7 and they will kill faster than humans can act to defend themselves."

Governments are meeting at the UN in Geneva on Monday for the fifth time to discuss whether and how to regulate lethal autonomous weapons systems (Laws). Also known as killer robots, these AI-powered ships, tanks, planes and guns could fight the wars of the future without any human intervention.

The Campaign to Stop Killer Robots, backed by Tesla's Elon Musk and Alphabet's Mustafa Suleyman, is calling for a preemptive ban on Laws, since the window of opportunity for credible preventative action is fast closing and an arms race is already in full swing.

"In 2015, we warned that there would be an arms race to develop lethal autonomous weapons," says Walsh. "We can see that race has started. In every theatre of war—in the air, on the sea, under the sea and on the land—there are prototype autonomous weapons under development."

Ahead of the meeting, the US has argued that rather than trying to "stigmatise or ban" Laws, innovation should be encouraged, and that use of the technology could actually reduce the risk of civilian casualties in war. Experts, however, fear the systems will not be able to distinguish between combatants and civilians and act proportionally to the threat.

Drones are one example of autonomous weapons that are already being used, including by the United States.

France and Germany have been accused of shying away from tough rules by activists who say Europe should be leading the charge for a ban. A group of non-aligned states, led by Venezuela, is calling for the negotiation of a new international law to regulate or ban killer robots. The group seeks general agreement from states that "all weapons, including those with autonomous functions, must remain under the direct control and supervision of humans at all times."

The New Global Arms Race

Fully autonomous weapons do not yet exist, but high-ranking military officials have said the use of robots will be widespread in warfare in a matter of years. At least 381 partly autonomous weapon and military robotics systems have been deployed or are under development in 12 states, including China, France, Israel, the UK and the US.

Automatic systems, such as Israel's Iron Dome and mechanised sentries in the Korean demilitarised zone, have already been deployed but cannot act fully autonomously. Research by the International Data Corporation suggests global spending on robotics will double from $91.5bn in 2016 to $188bn in 2020, bringing full autonomy closer to realisation.

The US, the frontrunner in the research and development of Laws, has cited autonomy as a cornerstone of its plan to modernise its army and ensure its strategic superiority across the globe. This has caused other major military powers to increase their investment in AI and robotics, as well as in autonomy, according to the Stockholm International Peace Research Institute.

"We very much acknowledge that we're in a competition with countries like China and Russia," says Steven Walker, director of the US Defense Advanced Research Projects Agency, which develops emerging technologies and whose 2018 budget was increased by 27% last year.

The US is currently working on the prototype for a tail-less, unmanned X-47B aircraft, which will be able to land and take off in extreme weather conditions and refuel in mid-air. The country has also completed testing of an autonomous anti-submarine vessel, Sea Hunter, that can stay at sea for months without a single person onboard and is able to sink other submarines and ships. A 6,000kg autonomous tank, Crusher, is capable of navigating incredibly difficult terrain and is advertised as being able to "tackle almost any mission imaginable."

The UK is developing its own unmanned vehicles, which could be weaponised in the future. Taranis, an unmanned aerial combat vehicle drone named after the Celtic god of thunder, can avoid radar detection and fly in autonomous mode.

Russia, meanwhile, is amassing an arsenal of unmanned vehicles, both in the air and on the ground; commentators say the country sees this as a way to compensate for its conventional military inferiority compared with the US. "Whoever leads in AI will rule the world," said Vladimir Putin, the recently re-elected Russian president, last year. "Artificial intelligence is the future, not only for Russia, but for all humankind."

Russia has developed a robot tank, Nerehta, which can be fitted with a machine gun or a grenade launcher, while its semi-autonomous tank, the T-14, will soon be fully autonomous. Kalashnikov, the Russian arms manufacturer, has developed a fully automated,

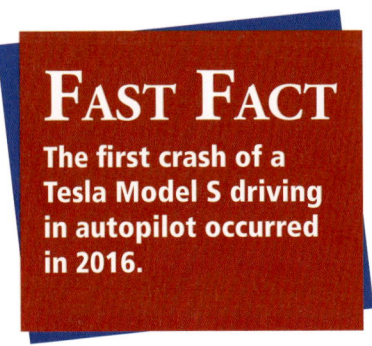

Fast Fact
The first crash of a Tesla Model S driving in autopilot occurred in 2016.

high-calibre gun that uses artificial neural networks to choose targets.

China has various similar semi-autonomous tanks and is developing aircraft and seaborne swarms, but information on these projects is tightly guarded. "As people are still preparing for a high-tech war, the old and new are becoming intertwined to become a new form of hidden complex 'hybrid war,'" wrote Wang Weixing, a Chinese military research director, last year.

"Unmanned combat is gradually emerging. While people have their heads buried in the sand trying to close the gap with the world's military powers in terms of traditional weapons, technology-driven 'light warfare' is about to take the stage."

Pandora's Box

According to the Campaign to Stop Killer Robots, these systems threaten to become the "third revolution in warfare," after the invention of gunpowder and nuclear bombs, and once the Pandora's box is opened it will be difficult to close.

"Inanimate machines cannot understand or respect the value of life, yet they would have the power to determine when to take it away," says Mary Wareham, the campaign coordinator. "Our campaign believes that machines should never be permitted to take human life on the battlefield or in policing, border control, or any circumstances."

Supporters of a ban say fully autonomous weapons are unlikely to be able to fully comply with the complex and subjective rules of international humanitarian and human rights law, which require human understanding and judgment as well as compassion.

Pointing to the 1997 ban on landmines, now one of the most widely accepted treaties in international law, and the ban on cluster munitions, which has 120 signatories, Wareham says: "History shows how responsible governments have found it necessary in the past to supplement the limits already provided in the international legal framework due to the significant threat posed to civilians."

It is believed that the weaponisation of artificial intelligence could bring the world closer to apocalypse than ever before. "Imagine swarms of autonomous tanks and jet fighters meeting on a border and one of them fires in error or because it has been hacked," says Noel Sharkey, professor of artificial intelligence and robots at the University of Sheffield, who first wrote about the reality of robot war in 2007.

"This could automatically invoke a battle that no human could understand or untangle. It is not even possible for us to know how the systems would interact in conflict. It could all be over in minutes with mass devastation and loss of life."

> **EVALUATING THE AUTHOR'S ARGUMENTS:**
>
> In this viewpoint Mattha Busby analyzes the topic of futuristic autonomous weapon systems. Using examples from the viewpoint, what are the pros and cons of such systems? Do you think it's possible to effectively regulate autonomous weapons?

Viewpoint 5

Privacy Laws Need to Change in the Era of Big Data

"The 2008 Illinois Biometric Information Privacy Act was one of the first laws to attempt to regulate biometric privacy."

Danica Sergison

In the following viewpoint, Danica Sergison reports on the issue of privacy in the era of big data collection, particularly focusing on facial recognition technology and efforts to regulate biometric technology. Biometric technology is the use of technology to recognize a person based on a biological characteristic. Sergison demonstrates how difficult it is for private individuals to fight back against companies such as Google that use data without consent. She asserts that a better enforcement framework is necessary to ensure that privacy laws will be effective. Danica Sergison is a lawyer and peer support worker with an interest in tech. She is based in Toronto.

AS YOU READ, CONSIDER THE FOLLOWING QUESTIONS:
1. Why is it difficult to prove biometric information privacy violations, according to the author?
2. What are some of the risks associated with privacy violation as outlined in this viewpoint?
3. What suggestion does the author offer to combat this problem?

"Hard to prove harm: Google wins lawsuit over facial recognition," by Danica Sergison, Private Internet Access, January 6, 2019. https://www.privateinternetaccess.com/blog/2019/01/hard-to-prove-harm-google-wins-lawsuit-over-facial-recognition/. Licensed under CC BY-SA 4.0 International.

Facial recognition technology is a common feature of smartphones and social media, but this technology has the potential to be abused and misused.

As new privacy laws attempt to address the different ways that companies collect, store and use biometric data, it's also important to keep an eye on how the courts are interpreting and applying legislation.

In a recent court case, a federal judge ruled against a claim that Google had violated Illinois privacy laws by using uploaded pictures to create "face templates" without an individual's consent. The plaintiffs in the case either uploaded photos to an Android phone, or had photos uploaded of themselves. The photos were analyzed using Google's facial recognition software, which created a face template that allegedly was used to recognize individuals' personal characteristics, including age, gender, race and location.

As reported by Courthouse News, the 2008 Illinois Biometric Information Privacy Act was one of the first laws to attempt to regulate biometric privacy. The law is also unique in that it allows private individuals to sue for damages when their rights under the law are violated, including as a class action. Civil remedies like this can provide a way for individuals to be compensated for harms and losses, as well as provide an additional financial incentive for

> **FAST FACT**
> "Selfie" was named the word of the year by Oxford Dictionaries in 2013.

companies to respect individual privacy rights.

Substantial Risk

The challenge with civil remedies is that they generally require a demonstration of harm or injury—and not all types of harm or injury are recognized by courts, or easy to prove. Privacy violations fall into this category, and the text of the decision illustrates some of the reasons why.

While Google collected, stored and applied facial recognition processes to individuals' photos without consent, part of the reason the legal case failed was because the plaintiffs couldn't demonstrate that they had actually been harmed. They argued that with their photographs and related personal information could be sold, used internally for profit or advertising, or inadvertently compromised in a data breach.

However, the judge found that even though the information was gathered without consent, no actual harm had occurred. It wasn't enough that the plaintiffs feared that their information would be compromised, used to profit, or sold in the future—that harm had to actually happen, or be substantially likely to happen. Evidence of recent breaches at Google (Google+) and potential future uses of the technology was not enough to persuade the judge that the potential for harm was high enough for the claim to proceed.

Privacy Laws Need Effective Enforcement Tools

Cases like this illustrate some key areas of difficulty in drafting privacy legislation. While harm doesn't always need to be proven for the state to impose a fine or other penalty, cases brought by individuals are generally held to a different standard. In a world where privacy breaches can have swift and dramatic consequences for personal health and safety, we need enforcement tools that can address risk proactively, rather than waiting until the damage is done.

In the meantime, maybe it's time to delete those selfies from Google Photos.

EVALUATING THE AUTHOR'S ARGUMENTS:

In this viewpoint Danica Sergison reports on an issue that affects many social media users. She suggests that large companies like Google can use a person's data without their consent. Do you think that collecting data like age, gender, and race through facial recognition is a major breach of privacy? Why or why not?

Viewpoint 6

Should the Rights of Smart Robots Be Protected?

E. Van Trigt

"Sophisticated robots, i.e. sentient AI robots, should receive a new legal status, the 'electric person' in order to ensure their rights and responsibilities."

In this viewpoint, E. Van Trigt writes about what some might consider a far-flung topic: the need to grant robots rights. Van Trigt reports that some experts see the need for a legal framework seeking to outline rights for smart robots. Van Trigt provides opinions from some experts that there will come a time when smart, sentient robots will be among us, and their rights should be protected. E. Van Trigt holds a master's degree in public international law and philosophy and writes for Peace Palace Library, an international law library based in the Hague.

AS YOU READ, CONSIDER THE FOLLOWING QUESTIONS:
1. Which country was the first to grant citizenship to a robot, according to Van Trigt?
2. According to this viewpoint, what is a strong AI?
3. As reported by Van Trigt, what is the meaning of "second-order rights?"

"Robots and Rights," by E. Van Trigt, MA in public international law and in philosophy of law, Peace Palace Library, November 16, 2017. Reprinted with permission.

Humans are granted various legal rights, and some have begun to question whether smart robots and AI should be granted the same rights.

Recently, an intelligent and human looking robot named Sophia made global headlines when Saudi Arabia granted the humanoid robot Saudi citizenship. According to the headlines, Saudi Arabia became the first country to grant a robot citizenship. The news caused quite a stir—the female looking robot was not wearing a hijab, she was not accompanied by a male guardian and the robot was awarded citizenship, which made it look like a humanoid intelligent robot was given more rights than women and migrants living in Saudi Arabia. Saudi Arabia is known for its restrictive policy concerning women's rights and Saudi women have only recently been given the right to drive a car.

Publicity Stunt

Perhaps it is a bit premature to give an AI humanoid robot like Sophia citizenship rights. Was it a publicity stunt? Yes. Robot Sophia was created by Hanson Robotics in Hong Kong. She appeared on stage at the Future Investment Initiative in Saudi Arabia on Wednesday 25 October. Although robot Sophia was said to be "basically alive,"

How Should AI and Robots Be Regulated to Benefit the Future of the Human Race?

> **FAST FACT**
> Sophia the humanoid robot was interviewed on the talk shows *60 Minutes*, *Good Morning Britain*, and the *Tonight Show with Jimmy Fallon*.

in fact she is not a very sophisticated AI robot; being neither sentient nor self-aware.

People that were able to meet, greet and interview Sophia, concluded that her answers to open-ended questions seemed to be programmed responses. Humanoid AI robot Sophia was designed to grow smarter over time and is able to answer questions and recognise voices and faces. She has been created as a companion for elderly persons at nursing homes, but for example she also could be used as an assistant at large events.

Increasingly robots and AI are becoming part of every day life. It won't be long until intelligent robots like Sophia or more advanced humanoid robots will serve as companions and or assistants in many households. The use of these sophisticated AI robots have not been properly regulated yet but fortunately the world is becoming more and more aware of the importance of creating ethical guidelines and legislation concerning the manufacturing and use of sophisticated AI robots.

In recent international law literature especially one type of robots is often being discussed: an autonomous weapons system which is often used as an instrument for targeted killing. In this regard, UN's Special Rapporteur on extrajudicial, summary or arbitrary executions, Mr. Christof Heyns stated in a report (2013) that military lethal autonomous robotics can not fully comply with the laws of war. These robots are autonomous but this has nothing to do with the notions of free will, human judgement, and moral autonomy. Special Rapporteur Mr. Heyns thinks that truly sentient AI robots or strong AIs are "not currently in the picture." But there are many experts who think that in a few decades significant technological developments could lead to truly intelligent AI (humanoid) robots.

The Need for a Legal Framework

For decades literature and sci-fi movies have warned humankind for the dangers posed by robots and AI. Many sci-fi books, movies and TV series have portrayed, in several disturbing dystopic scenarios, that things can seriously go wrong when the use of robot technology and AI is not guided properly and consciously within a moral and legal framework. In 1950, the science fiction writer Isaac Asimov introduced the three laws of robotics in his work: *I, Robot*. These laws basically serve to ensure that an intelligent robot should never be used for purposes which are destructive for humankind.

On the invitation of the Committee on Culture, Science, Education and Media of the Parliamentary Assembly of the Council of Europe (PACE), the Dutch Rathenau Institute conducted research concerning challenges arising from the use of robotics, artificial intelligence, and virtual and augmented reality. The research focused on issues relating to the right to respect private life, human dignity, ownership, safety and liability, freedom of expression and the prohibition of discrimination as well as access to justice and the right to a fair trial. The institute issued a publication in which it proposed two new human rights: the right to not be measured, analyzed or coached, and the right to meaningful human contact.

In January 2017, the European Parliament made a recommendation to the EU Commission concerning a proposal that sophisticated robots, i.e. sentient AI robots should receive a new legal status, the "electric person" in order to ensure their rights and responsibilities. Rapporteur Mady Delvaux MEP, stated: "A growing number of areas of our daily lives are increasingly affected by robotics. In order to address this reality and to ensure that robots are and will remain in the service of humans, we urgently need to create a robust European legal framework."

In the above mentioned proposal which is concerned with the civil law rules on robotics, several recommendations are made, such as:

- A call on the European Commission to propose a legal definition of "smart autonomous robots and their subcategories with a system of registration of the most advanced of them."
- "A guiding ethical framework for the design, production and use of robots."
- "Creation of a European Agency for robotics and artificial intelligence."

- Asks the European Commission to look further into issues regarding intellectual property rights and the flow of personal data (privacy and data protection related matters).
- A call for the "international harmonisation of technical standards," and of European legislation concerning robots and AI.
- The need for uniform criteria which can be used to identify in which areas experiments with robots are permitted.
- Civil liability related issues, such as the proposal of a new mandatory insurance scheme which concerns the coverage of damage caused by company robots.

Smart AI Robots and Human Rights

Even though the robot Sophia was recently made a citizen of Saudi Arabia, it is perhaps a bit premature to talk about granting rights to smart robots. Real sentient, smart AI robots do not exist yet, but many already fear the idea of giving rights to robots.

The question whether a sentient AI robot deserves rights or should have rights is a heavily debated topic and raises many questions. Human rights have already been extended to animals and corporations. Granting rights to intelligent AI robots could lead to many unforeseen consequences.

If in the future sophisticated, sentient and self-aware AI robots would be invented and would be granted human rights, this would put them on equal footing with humans. As bearers of rights, these AI robots would have the right to vote, the right to self-determination, the right to have possessions, the right to conclude legal contracts and have the right not to be anyone's property. Is this desirable? Is it wise to give an artificial being rights? Would this also mean that a robot has to pay taxes and claim the right to self-determination? Intelligent Robots that are bearers of rights could reject to being used as a service robot.

There are experts who really believe that technological developments concerning robotics will eventually result in the need for new legal and ethical guidelines related to the use of smart robots. Scholar Dr. Kate Darling from MIT Media Lab for example is an advocate of second-order rights for robots. This type of rights serves as protection to prevent the mistreatment of sophisticated AI robots. Second-order rights could be compared to animal protection laws and guidelines which are created in order to prevent animal abuse and inhumane treatment. The question is whether inanimate beings such as intelligent robots could be compared with animals when it concerns their legal status. While some of these problems with intelligent robots have been extensively portrayed in sci-fi literature and movies, this still seems unrealistic, far-fetched and even preposterous to many.

Creating a legal status of the electric person as proposed by the EU is a bold initiative and in the eyes of many perhaps also a bit precocious. However, it is wise and necessary to further investigate the dangers, advantages and (legal) consequences of the development and use of intelligent AI robots and to map out the necessary conditions for an adequate legal framework for robotics and AI.

As far as human rights for robots are concerned, besides the question whether it is desirable and reasonable to give smart AI robots rights, one should not forget that granting "rights" to robots could also have many unforeseen and unwanted consequences. More in depth legal research in this field is therefore also necessary.

EVALUATING THE AUTHOR'S ARGUMENTS:

This viewpoint suggests that some people believe in granting legal rights to smart robots, while others see it as preposterous. Which side are you on? Explain your reasoning. What variables should be considered when determining who or what to grant legal rights?

Facts About AI, Robots, and the Future of the Human Race

Editor's note: These facts can be used in reports to add credibility when making important points or claims.

Important Concepts and Definitions

- **artificial intelligence:** A branch of computer science dealing with the simulation of intelligent behavior in computers.
- **human-robot interaction:** A field of research that studies relationships between humans and robots.
- **multiplicity:** The idea that AI and robots will not take over but work with and alongside humans.
- **robot:** A machine that can perform autonomous tasks and has the ability to use senses and manipulate its environment.
- **The Singularity:** The point in time where machines take over and humans are forced into subservience or out of existence.

Important People in the Field of AI

- **Isaac Asimov:** Well-regarded science fiction writer who came up with three laws pertaining to the development of robots:
  - A robot may not injure a human being or allow a human to come to harm.
  - A robot must obey orders given to it by a human except when those orders would conflict with the first law.
  - A robot must protect its existence as long as this does not conflict with either the first or second law.
- **Bill Gates:** Founder of Microsoft who feels AI can be both beneficial and dangerous.
- **Stephen Hawking:** Theoretical physicist who was an outspoken critic of AI.

- **John McCarthy:** In 1955, he claimed that "every aspect of learning or any other feature of intelligence can in principle be so precisely described that a machine can be made to simulate it."
- **Elon Musk:** Engineer and CEO of Space-X who is worried about the negative potential of AI.
- **Alan Turing:** Computer scientist and mathematician known for breaking Nazi codes and developing the Turing test.

Important Dates Pertaining to Robots:

1956 The word "robot" is used for the first time.

1960s Silicon Valley developed its first robot called Shakey.

1980s Honda makes a humanoid robot called P3.

2016 A Tesla Model S driving in autopilot crashes and kills its owner.

Organizations to Contact

The editors have compiled the following list of organizations concerned with the issues debated in this book. The descriptions are derived from materials provided by the organizations. All have publications or information available for interested readers. The list was compiled on the date of publication of the present volume; the information provided here may change. Be aware that many organizations take several weeks or longer to respond to inquiries, so allow as much time as possible for the receipt of requested materials.

All Turtles Corporation
1266 Harrison St.
San Francisco, CA 94103
website: www.all-turtles.com
An up-and-coming technology start up, this company is on the cutting edge of AI. Read their newsletter, watch videos, and listen to podcasts to get the latest information on AI trends.

Brookings Institution
1775 Massachusetts Ave. NW
Washington, DC 20036
phone: (202) 797-6000
website: www.brookings.edu
The Brookings Institution is a nonprofit think tank that draws on the expertise of over 300 experts in numerous fields. Its information spans national and international articles and print materials pertaining to artificial intelligence and emerging technologies.

The Conversation
89 South St., Suite 202
Boston, MA 02111
website: www.theconversation.com/us
The Conversation is an independent global site dedicated to sharing

news important to people around the world. They believe that democracy depends on knowledge of the news. Read articles on a wide variety of topics including AI, robotics, and technology.

Interesting Engineering
website: www.interestingengineering.com
This engineering site gives the latest news on the developments of AI and robotics. Read articles or watch videos on the newest science, technology, and invention.

Machine Intelligence Research Institute (MIRI)
2036 Bancroft Way
Berkeley, CA 94704
email: contact@intelligence.org
website: www.intelligence.org
This organization supports research to ensure that smarter-than-human AI will have a positive impact. Read current articles and subscribe to the newsletter to keep up-to-date on their research.

Pew Research Center
1615 L St. NW, Suite 800
Washington, DC 20036
phone: (202) 419-4300
email: info@pewresearch.org
website: www.pewresearch.org
The Pew Research Center conducts surveys and research in order to inform the public about the issues, attitudes, and trends that are important and inform the world. This site contains articles about technology, AI, and robotics, which include surveys about how people see these issues impacting their own lives.

UnDark Magazine
77 Massachusetts Ave., E19-623
Cambridge, MA 02139
phone: (617) 230-8456
email: info@undark.org
website: www.undark.org

Read short articles, investigate longer case studies and reports, and listen to podcasts on a wide variety of science-based information.

World Economic Forum
350 Madison Ave.
New York, NY 10017
phone: (212) 703-2300
website: www.weforum.org
Inform yourself on topics of cybersecurity, AI, technology, robotics, and more by reading and watching videos on this comprehensive website.

For Further Reading

Books

Challoner, Jack. *Artificial Intelligence.* New York, NY: DK Publishing, 2002. Important scientific concepts related to AI are presented in clear wording and accompanied by illustrations.

Cunningham, Anne C. *Artificial Intelligence and the Technological Singularity.* New York, NY: Greenhaven Publishing, 2017. This nonfiction anthology considers the controversies, ethical considerations, and potential threats involved in AI.

Ford, Martin R. *Rise of the Robots: Technology and the Threat of a Jobless Future.* New York, NY: Basic Books, 2015. This book examines the impact AI could have on economic prospects and on society as a whole by making many jobs performed by humans obsolete.

Kaplan, Jerry. *Artificial Intelligence: What Everyone Needs to Know.* Oxford, UK: Oxford University Press, 2016. An entrepreneur and AI expert helps explain the potential uses of AI in everyday life.

McPherson, Stephanie Sammartino. *Artificial Intelligence: Building Smarter Machines.* Minneapolis, MN: Twenty-First Century Books, 2018. This title examines the issue of machine intelligence, considering what it would mean for a computer to be able to think, what the potential dangers are, and whether consciousness and self-awareness are possible for machines.

Mindell, David A. *Our Robots, Ourselves: Robots and the Myth of Autonomy,* New York, NY: Viking, 2015. An MIT professor explains the relationship between humans and machines and debunks common myths about robotics.

Oppenheimer, Andres. *The Robots Are Coming! The Future of Jobs in the Age of Automation.* New York, NY: Vintage Books, 2019. Through extensive research and numerous interviews with experts, the author examines how various fields and areas of the workforce are being impacted by increases in automation and technological advances.

Otfinoski, Steven. *Making Robots: Science, Technology, Engineering.* New York, NY: Children's Press, 2017. This book explains how various kinds of robots are designed and constructed.

Smibert, Angie. *Artificial Intelligence: Thinking Machines and Smart Robots with Science Activities for Kids.* White River Junction, VT: Nomad Press, 2018. This title offers young readers a better understanding of what AI is and how it is developed. It includes numerous interactive activities to promote critical analysis.

Periodicals and Internet Sources

Artavia, Giovanni, "Artificial Intelligence in Future Society," Medium, April 7, 2019. https://medium.com/@giovaartavia/artificial-intelligence-in-future-society-23861f020e.

Atherton, Kelsey D, "Are Killer Robots the Future of War?" *New York Times*, November 5, 2018. https://www.nytimes.com/2018/11/15/magazine/autonomous-robots-weapons.html.

Blommaert, Jan, "We're All Data Subjects Now: The Citizen in the European GDPR," *Diggit Magazine*, May 21, 2018. https://www.diggitmagazine.com/column/were-all-data-subjects-now-citizen-european-gdpr.

Bode, Ingvild, "AI Has Already Been Weaponized—and It Shows Why We Should Ban Killer Robots," *Conversation*, September 6, 2018. https://theconversation.com/ai-has-already-been-weaponised-and-it-shows-why-we-should-ban-killer-robots-102736.

Callahan, Molly, "Should Robots Have Rights?" PHY.org, December 8, 2017. https://phys.org/news/2017-12-robots-rights.html.

Clairmont, Nicholas, "How Close Is the Turing Test to Being Beaten?" *BIG THINK*, June 20, 2012. https://bigthink.com/humanizing-technology/how-close-is-the-turing-test-to-being-beaten.

Clifford, Catherine, "Elon Musk: Mark My Words—AI is Far More Dangerous Than Nukes," CNBC News, March 13, 2018. https://www.cnbc.com/2018/03/13/elon-musk-at-sxsw-a-i-is-more-dangerous-than-nuclear-weapons.html.

Dashevsky, Evan, "Do Robots and AI Deserve Rights," *PC*, February 16, 2017. https://www.pcmag.com/article/351719/do-robots-and-ai-deserve-rights.

Driessen, Lennart, "Google and Big Data: Are We Heading Towards a Dystopian Reality?" *Diggit Magazine,* January 4, 2018. https://www.diggitmagazine.com/articles/google-and-big-data-are-we-heading-towards-dystopian-reality.

Friend, Tad, "How Frightened Should We Be of AI?" *New Yorker,* May 7, 2018. https://www.newyorker.com/magazine/2018/05/14/how-frightened-should-we-be-of-ai.

Griffiths, Sarah, "Do You Fear AI Taking Over?" *Daily Mail,* March 11, 2016. https://www.dail?ymail.co.uk/sciencetech/article-3487851/Do-fear-AI-taking-people-believe-computers-pose-threat-humanity-fear-ll-steal-jobs.html.

Hsu, Jeremy, "Forget Killer Robots: Autonomous Weapons Are Already Online," *UnDark,* July 9, 2018. https://undark.org/article/killer-robots-autonomous-weapons-online/.

Jones, Rory-Cellan, "Stephen Hawking Warns Artificial Intelligence Could End Manking," BBC News, December 2, 2014. https://www.bbc.com/news/technology-30290540.

Muehlhauser, Luke, "What Is AGI," Machine Intelligence Research Institute, August 11, 2013. https://intelligence.org/2013/08/11/what-is-agi/.

Palmer, Annie, "Most People Now Think Artificial Intelligence Poses a Threat to the Human Race, Study Claims," *Daily Mail*, April 25, 2019. https://www.dailymail.co.uk/sciencetech/article-6960317/Most-people-think-artificial-intelligence-poses-threat-human-race-study-claims.html.

Rawlinson, Kevin, "Microsoft's Bill Gates Insists AI Is a Threat," BBC News, January 29, 2015. https://www.bbc.com/news/31047780.

Ryder, Mark, "Killer Robots Already Exist, and They Have Been Here a Very Long Time," *Conversation*, March 27, 2019. https://theconversation.com/killer-robots-already-exist-and-theyve-been-here-a-very-long-time-113941.

Starr, Michelle, "10 Facts About the Apple-1, the Machine that Made Computing History," CNET, June 29, 2016. https://www.cnet.com/news/apple-1-the-machine-that-made-computing-history/.

Websites

CODE (code.org)

Kids of all ages can learn to code on this site. It provides users with the tools to construct working apps, websites, and games by using CSS, HTML, JavaScript, and more.

National Geographic: Challenge Robots (www.nationalgeographic.org/interactive/challenge-robots/)

Young people can take on the role of an engineer specializing in robotics. Learn the parts of a robot and basic robotics programming, then design a robot to solve a real-world problem.

SCRATCH (scratch.mit.edu)

SCRATCH is a website where young people can learn to program their own games, animations, and interactive stories. They can also share their creations with other people as part of the online community.

Index

A
Alexa, 16, 75
algorithms, 9, 13, 16, 24, 49, 58, 68, 78, 90
Apple Computers, 8, 16, 20, 40
applications, AI, 15, 68
artificial general intelligence (AGI), 18, 19, 21, 22, 26, 28
artificial intelligence (AI)
 application of, 12, 13, 15–16, 28, 31, 33, 36, 43, 52, 55–60, 68, 71, 75, 78
 definition of, 7–8, 13, 19, 24, 25, 26, 27, 44, 59
 malicious use of, 24, 43, 44, 45–46, 48, 49–53, 55, 61, 62, 68, 75
artificial intelligence (AI) systems, 18, 19, 24, 28, 37, 41, 48, 49, 50, 51
artificial neurons, 14
Asimov, Isaac, 31, 62, 105, 108
Avent, Ryan, 81, 86

B
Benson, Greg, 74, 75, 80, 91
Berlin, Heather, 26, 27
biometric technology, 98, 99
Bletchley Park (United Kingdom), 38
Bowcott, Owen, 87, 91
brain-machine interfaces, 24, 26

brains, human, 14, 19, 20, 21, 22, 23, 24, 25, 26, 27, 28, 64, 77
Busby, Mattha, 92, 97

C
Campaign to Stop Killer Robots, 93, 96
Čapek, Karel, 31, 62
chatbots, 16, 38, 40
communication, 26, 33, 38, 40, 49
consciousness, 23, 24, 26–27, 28, 44, 79, 105

D
deep learning, 12, 15, 20, 24, 32, 35, 55, 56, 57, 58, 70
Dickinson, Dave, 61, 66

E
emotions, 16, 23, 24, 26, 61, 65, 77, 79

F
facial recognition, 55, 98–99, 100, 101, 104
factory automation, 30, 36, 62, 89

G
Gasparetto, Alessandro, 30, 31, 36
Google, 16, 20, 98, 99, 100, 101
Google Assistant, 16, 75
Google China, 40

H
Harris, Sam, 43, 44–47
Hawking, Stephen, 7, 8, 43, 45, 46, 47, 64, 75, 108
Herget, Steffen, 12, 17
Herweijer, Celine, 67, 72
Husain, Amir, 26, 27, 28

I
IBM, 13, 16, 57
imitation, 18, 38, 39
Intelligent Virtual Assistant (IVA) software, 46

J
jobs lost due to automation, 8, 9, 17, 64, 68, 80, 81, 83, 84, 87, 88, 89, 90

K
killer robots, 55, 92, 93, 94, 96. *See also* robots, malevolent
Kutzweil, Ray, 23, 24

L
language, 16, 20, 34, 40, 75, 91
Laws of Robotics, 31, 62, 105
lethal autonomous weapons systems (Laws), 93, 95
locomotion, 21–22. *See also* movement
Loebner Prizes, 38, 40
logical reasoning, 13, 32, 33, 35

M
machine learning, 12, 15, 16, 18, 19, 20, 24, 38, 52, 55, 56, 75, 78, 79

Malicious Use of Artificial Intelligence, 48, 49
McCarthy, John, 7, 90, 109
Michaels, Vivian, 43, 47
movement, 26, 33, 34, 35, 40. *See also* locomotion
Musk, Elon, 8, 62, 65, 93, 109

N
neural AI, 12, 14, 15
neural implants, 26
neural networks, 14, 15, 16, 27, 96
neural processing units (NPUs), 16
nuclear weapons, 46, 62

O
One Hundred Year Study on Artificial Intelligence (AI100), 25
opinion analysis, 16, 51

P
patterns and pattern detection, 13, 15, 24, 61, 90
photographs, pictures, and images, 6, 14, 15, 20, 21, 37, 38, 51, 99, 100
Prasanna, Narasimha, 18, 19, 22
privacy, 9, 98, 99, 100, 101, 106

R
Ramirez, Vanessa Bates, 23
reason, ability to, 9, 32, 35
recognition, 15, 16, 20–21, 33, 37, 38, 55, 98–99, 104

reinforcement learning, 15, 18, 21–22, 70, 71
rights of humans, 9, 89, 90, 96, 99, 100, 103, 105, 106–107
rights of robots, 9, 62, 102–107
robotics, 22, 30, 31, 32, 35, 36, 59, 65, 70, 81, 83, 84, 88, 89, 94–95, 104, 105, 106, 107
robots, 7, 8, 9, 43, 49, 56, 59, 61, 82, 87, 89, 90, 91, 104, 106, 107
   definition of, 31, 105
   fear of, 65–66
   industrial, 30, 31, 32–36, 64, 83, 84
   malevolent, 43, 44, 55, 58, 62, 78, 92, 93, 94, 96, 97

S
Scalera, Lorenzo, 30, 31, 36
self-awareness, 27, 33, 40, 77, 104, 106
self-driving cars, 28, 38, 40, 49, 52, 68, 70, 75, 77
sensation, 9, 14, 27, 34, 108
sentiment analysis. *See* opinion analysis
Sergison, Danica, 98, 101
Singularity, the, 65, 108
Siri, 16, 64, 75
Smith, Rob, 48, 49, 53
social media, 9, 51, 70, 101
Sophia (robot), 103, 104, 106
Space-X, 8, 109
speech processing, 37, 38
Stone, Peter, 24–26, 28–29, 49–50

supercomputers, 13, 19, 57, 70
supervised learning, 15, 18, 19, 20, 21, 22
symbolic AI, 12, 13, 14, 15, 16
synthesized speech, 16, 40

T
tasks, execution of, 8, 21, 24, 26, 28, 31, 33, 34, 35, 46, 50, 56, 82, 83, 108
Turing, Alan, 13, 37, 38, 39, 40, 41, 62, 109
Turing test, 13, 27, 37, 38–40, 41, 62, 109

U
"Uncanny Valley," 61, 65–66

V
Van Trigt, E., 102
Vyas, Kashyap, 54

W
Wallach, Wendell, 51, 52
Watson (supercomputer), 16–17
weapons, 46, 58, 62, 93, 96
   autonomous, 9, 45–46, 68, 92, 93, 94, 95, 97, 104
   intelligent, 19, 90, 97
weather forecasting, 57, 58, 67, 70, 71
World Economic Forum, 51, 52, 68
Worswick, Steve, 38–39
Wozniak, Steve, 8, 40

Z
Zerega, Blaise, 37, 38, 41

Picture Credits

Cover the Asahi Shimbun/Getty Images; p. 11 Anton Gvozdikov/Shutterstock.com; p. 14 © AP Images; p. 20 snapgalleria/Shutterstock.com; p. 25 Rick Friedman/Corbis Historical/Getty Images; p. 32 Betastock/Shutterstock.com; p. 39 Science History Images/Alamy Stock Photo; p. 42 MAD.vertise/Shutterstock.com; p. 45 Jeff Bukowski/Shutterstock.com; p. 50 kung_tom/Shutterstock.com; p. 55 Chip Somodevilla/Getty Images; p. 63 AF archive/Alamy Stock Photo; p. 69 Macrovector/Shutterstock.com; p. 73 Sakaret/Shutterstock.com; p. 76 VDB Photos/Shutterstock.com; p. 82 Syda Productions/Shutterstock.com; p. 88 tsyhun/Shutterstock.com; p. 94 Glenn Price/Shutterstock.com; p. 99 Maskot/Getty Images; p. 103 wavebreakmedia/Shutterstock.com.

Photo Researcher: Sherri Jackson